The Missing Ingredient

The Missing Ingredient

The Curious Role of Time in Food and Flavor

JENNY LINFORD

The Overlook Press
New York, NY

To my family and friends,
with my love

This edition first published in the United States in 2018
by The Overlook Press, Peter Mayer Publishers, Inc.

NEW YORK
141 Wooster Street
New York, NY 10012
www.overlookpress.com

For bulk and special orders, please contact sales@overlookny.com,
or write us at the above address.

Cataloging-in-Publication Data is available from the Library of Congress

Manufactured in the United States of America

ISBN: 978-1-4683-1638-4

2 4 6 8 10 9 7 5 3 1

Contents

Introduction 1

SECONDS 11

MINUTES 45

HOURS 139

DAYS 189

WEEKS 239

MONTHS 277

YEARS 317

Food Heritage: The Work of Centuries 351

Select Bibliography 357

Food and Drink Directory 359

Acknowledgements 363

Index 365

Introduction

The idea for this book came to me in the early hours – between 4 a.m. and 6 a.m. – when night passes to day. It's a strange, solitary period when I, as a light sleeper, often find myself lying half-awake in the dark in bed. As I lie there, thoughts – often worries, but also solutions and ideas – drift up from the depths of my subconscious to the surface of my mind. I was preoccupied, pondering ideas for a food-themed day to offer as an event. *Could I take people on a journey through food history?* I wondered ambitiously. Travel from the early days of food in Britain through different periods in time . . . Time. The word caught me, stopping me in my mental tracks. My brain took the word and spun it around – rather than food in time, what about time in food; time as an ingredient, THE universal, invisible ingredient. It felt like a eureka moment, an insight into something big. Rather than a theme for an event, I felt I had the kernel of an idea for a book.

This book explores culinary time. In over 25 years of writing and thinking about food, I have always been fascinated by the way in which food connects to so many aspects of our lives – history, identity, culture. Food is so much more than fuel for our bodies; it is charged with meaning. Time and food is a vast, complex subject and one which I find enthralling. I spent months working out a structure – what I wanted to put in the book; time touches all aspects of food in some way or the other, but I wanted a book that was engaging and interesting. I think of this book as a kaleidoscope, in which my short essays, each looking at time through the prism of an aspect of food or cooking, fall together to form a complex, intricate pattern. Writing *The Missing Ingredient* has been an engrossing journey, enlivened by encounters with remarkable people en route. During my encounters with food producers, chefs and food writers, the importance of taking enough time came up repeatedly. For the most part, this was without prompting from me. Very few of the people I talked to knew that it was culinary time I was focused on (a book about 'the craft of food, what makes good food' is how I usually described it), as I wanted to discern for myself the role that time played in their work. *The Missing Ingredient* moves from seconds to years, taking the reader on a journey with me through aspects of culinary time. The book is designed to be read from start to finish but also works as a volume to be dipped into. What I realized, as I researched and wrote, was the overlapping nature of the material. How looking at a subject – such as meat – from different time-focused points of view meshes together to create an insight into the complex, varied role of time: from the time required to rear livestock for maximum flavour, to hanging meat, to the right amount of time to cook it or the ingenious, historic ways we have found to preserve this valuable protein.

Time is the universal ingredient in the food we cook and eat. As an invisible ingredient, however, it is seldom considered in its own right. Time is an essential part of the act of cooking. To cook food well, one needs to know how to use time appropriately. Time in the kitchen can be a brisk affair measured in minutes: the quick cooking of spaghetti in boiling water until just al dente, the exciting, rapid transformation that happens when one plunges sliced potatoes into hot oil to make French fries, the thickening as you stir a mixture of flour, butter and milk into a sauce. It can also be altogether more mellow: gently simmering bones and vegetables for hours to make stock, marinating a shoulder of lamb with spices overnight, the steeping of Seville orange peel to make marmalade, slowly roasting a piece of pork belly at a low temperature for hours until the meat is succulent and the crackling crisp and golden-brown . . . When it comes to certain dishes – such as the frying of an omelette or the grilling of a fillet of fish – rapidity is of the essence in order not to overcook them. In these cases, a deft hand and an eye to the clock make all the difference between the creation of a meal that is pleasurable and one that is a lacklustre experience. Other aspects of cooking, however, require patience. The browning of meat – creating the appetizing aromas and flavours through the complex Maillard reaction – is a process that requires time in order to do it properly. In Malaysian cuisine, curries and sauces involve the initial cooking of a paste made from spices, herbs and flavourings. This must be done with care and cannot be rushed. Hastiness at this stage – a skimping of the time required – will manifest itself in a disappointing harshness and thinness of flavour in the final dish. Watch an experienced chef at work in the kitchen and one notices his or her acute awareness of time, how this factors into the decisions they are making. The ability to cook food for the 'right' amount of time is a key part of a cook's skills. One of the satisfactions of learning to cook is understanding how to use time effectively. Appropriately, a practical, working knowledge of culinary time takes time to acquire; there are no short-cuts to experience.

Time's relationship with food is multifaceted and complex, extending far beyond the right cooking times for dishes. We appreciate many different aspects of culinary time. At one end of the spectrum is the idea of freshness. Eating truly fresh food is a vivid experience. Years ago, when I was a teenager living in Italy, my parents and I visited Italian friends whose home was in the hills above Fiesole. Our kindly host, noticing my boredom at the adult conversation, told me that I could go into the garden and help myself to the fruit growing in it. I stepped out into the warm Tuscan sunshine and, rather hesitantly, reached up into the branches of an apricot tree, pulled a downy-skinned, deep-orange-red-coloured apricot off the tree and ate it. I still remember how delectable that apricot tasted. Not only was I eating it fresh from the tree – with no passage of time after harvesting for it to deteriorate – I was also eating it when it was ripe and ready. Over the months between spring and summer, a fertilized apricot blossom had been transformed, swelling and ripening with the rain and the warmth and light of the sun into the fruit that I enjoyed so much. The tree itself had taken years to grow, requiring care and attention from its owner from its first planting as a sapling to its fruitful maturity. My brief encounter with an apricot required many diverse types of time.

We also value the alluring patina that ageing adds to food and drink. Many of our most highly prized foods and drinks – Aceto Balsamico Tradizionale di Modena (traditional balsamic vinegar of Modena), *jamón ibérico*, vintage wines and spirits – take years to produce. This extensive use of time is closely bound up in our appreciation of these delicacies. The taking of time is nowadays used as a marketing tool. Recently, the multinational Unilever went to the trouble of producing a handsomely presented XO (Extra Old) Marmite. Labelled 'matured with a stronger taste', this is an example of ageing being used to add a sheen of glamour, transforming what is normally regarded as an everyday, mass-produced food into a desirable luxury.

So prized is the concept of ageing within the food world that numerous short-cuts are used to try to mimic its effects. Some cheese-makers tap into the cachet associated with matured cheeses by wrapping their cheeses in plastic, storing them in chilled conditions, then releasing them as 1- or 2-year-old cheeses. On tasting them, however, one would note very little difference or improvement. Meaningful cheese-maturing – such as that carefully carried out on Parmigiano-Reggiano or farmhouse Cheddars – allows the cheese to breathe and dry out. When consumed, these aged cheeses deliver a richness of umami and a long, lingering finish. An authentic ageing process is one that impacts meaningfully and beneficially on the final organoleptic (sensory) experience, so affecting aroma, taste and texture. There is a mysterious alchemy to how time works. The changes wrought by time during processes such as the maturing of air-dried meats or the ageing of port – even the simple melding of flavours in a stew made a day ahead – are little understood. The beneficial results, though, can be tasted and appreciated.

Our relationship with time and food is also a double-edged one. Time is the great destroyer, causing ingredients to deteriorate and decay. For centuries humans have sought to counteract the detrimental effects of time on precious, hard-won food by finding ways of preserving it. Drying, salting, smoking, fermenting, curing, pickling, saturating with sugar and – more recently – canning are among the ingenious ways developed to keep food safer for longer. With the advent of refrigeration and freezing, the battle to keep food fresh for longer made significant advances. While in many parts of the world the traditional methods of preserving food from decay are no longer strictly essential, we continue to use them, appreciating the way in which they enrich our diet.

Eating is essential for life. It is an act we repeat over and over again throughout our lives to sustain our bodies. Historically, it was understood that the preparation of food took time, from sourcing ingredients (through hunting, fishing, growing and trading) to the gutting, plucking, picking, kneading, pounding and grinding required before cooking could begin. Today, labour-saving devices, from bread-makers to food processors, make food preparation quicker and easier than ever before. Despite this, we seem to begrudge the time spent preparing and making food. The impatience that characterizes modern life is manifest in our approach to cooking.

'Cooking from scratch' is the term now given to what was simply called 'cooking' by my parents when I was a child; it specifically means starting with raw ingredients. For those who don't wish to make food in this way there are now numerous options. Since its invention in America in the 1950s, the ready-meal has moved from being an exotic novelty to a ubiquitous staple. The advent of the microwave in domestic homes means that chilled and frozen meals can be quickly and effortlessly heated through. The Horsegate scandal in Europe in 2013, when it was revealed that meat-based foods were being adulterated with undeclared horse meat and pork, caused a dip in ready-meal sales. Since then, though, the ready-meal market has bounced back and it is generally regarded as a buoyant sector of food manufacturing. Indeed, in Britain a recent report from business intelligence provider Key Note forecast that the market value of the ready-meals market would grow by 15.7% between 2015 and 2019. For those looking to cook for themselves, there are numerous already prepared ingredients designed to take time out of the process. Grated cheese, chopped onion, bagged salad leaves – simple kitchen tasks that previous generations took for granted now present commercial opportunities.

Dining out is another area of gastronomy where time is being reduced. The convivial idea of a long, leisurely meal at a restaurant with friends – sitting and lingering over coffee at the end – seems to belong to another era. The brutal realities of high rents in large cities have seen the development of table-turning as a common practice in order to maximize turnover. Meals are served within strictly allocated time frames and diners are expected to then move on promptly, so freeing up space for the next sitting. 'What you pay for when you eat out in London isn't ingredients or staff,' a restaurateur told me, 'it's rent.' One innovative café chain – Ziferblat – charges customers 8 pence a minute, rather than charging them for the food and drink they consume. 'Everything is free except for the time you spend,' explains founder Colin Shenton. The idea of eating out itself has been speeded up. Rather than going to the trouble of actually visiting restaurants, a popular app-delivered trend sees dishes from restaurants brought to your own home. 'Your favourite restaurants delivered fast to your door' is the tagline for Deliveroo, its service undertaken by teams of couriers pedalling through city streets to deliver meals as quickly as possible. Founded in 2013, it has quickly become one of Europe's best-funded start-ups, showing the financial potential of the online take-away model.

The history of food production has seen the move from individual producers towards mass-manufacturing. This shift brought with it the squeezing out of time from the making of food. 'Time is money' in the world of industrial food production, and much resourcefulness has been spent in minimizing time in processes. A striking example of this is the Chorleywood process, invented in Buckinghamshire in the early 1960s. It radically reformulated the ingredients and the method of making bread. Lower-protein flour, fat, yeast and additives are combined through a few moments of violent, mechanical agitation in powerful mixers. The resulting dough rises rapidly, meaning that loaves of bread can be baked far more quickly than bread made traditionally, as it had been for centuries. Today the Chorleywood process is used to produce 80% of Britain's bread. The results of bread made this way are dispiriting: light, insubstantial in texture, unpleasantly claggy to eat. And yet, around the world, there are examples of traditional time-consuming processes mimicked with short-cut methods by industrial food producers: brine-injected bacon, plastic-wrapped 'aged' beef, fizzy beers created by adding CO_2 rather than using natural fermentation, 'smoked' cheese made with smoke-flavoured additives . . . The quest to speed up our relationship with food continues to this day. The Space Age idea of an instant 'food meal pill' continues to enthral. Soylent is a meal-replacement product, available in drink, bar and powder form. It was invented in California in 2013 by Rob Rhinehart, a young tech start-up entrepreneur who created it to save himself time and money. Realizing that there was interest in the idea, he and his partners sought crowd-funding; such was the interest that they raised a massive investment very quickly indeed. Sales of Soylent – a mixture of protein, carbohydrates, lipids and micronutrients – are increasing each year. Its sales pitch is focused on saving time: 'If you've ever wasted time and energy trying to decide what to eat for lunch, or been too busy to eat a proper meal – Soylent is for you.' In this vision of the world, taking the time to cook and eat is simply a waste of time. As someone who loves to cook and eat, this purely functional approach to food is in opposition to everything I hold dear. Writing this book brought home to me the importance of time in creating good food.

And yet we have also seen the rise of craft food producers. While many are using traditional methods of making food, there is also a willingness to experiment and innovate. The artisan food scene is an interesting place, characterized by makers' imagination and energy. A striking common feature of their approach to creating good food is the proper use of time. Bakers making real sourdough bread allow the dough to ferment slowly and gently in order to create flavour and texture in the resulting loaf. Smokeries skilfully apply a time-consuming process of salting and cold-smoking to transform fresh, perishable fish into a cured luxury. Cheese-makers use different amounts of time to help them create a huge variety, ranging from fresh, moist cheeses – ready to eat within days of being made – to large cheeses, set aside and carefully matured in the right conditions for months. Livestock farmers committed to producing high-quality meat understand very well the importance of time in rearing their animals. Slower growing times allow the development of intramuscular fat and collagen, strong bones filled with marrow; the reward for the consumer being meat or poultry that is succulent, tender, flavourful. The results are foods to be savoured and relished. They possess a power to jolt us, bring us to the present, make us think about what we are eating.

Young chefs are embracing time-consuming techniques such as curing meats, pickling and fermenting because they enjoy mastering these fundamental aspects of working with ingredients. The satisfaction of making food also seems to be rippling into people's homes, beginning with how food is sourced. There has been a rise of food markets and farmers' markets in cities and towns, offering a more leisurely, personal and seasonally connected way of shopping rather than simply pushing a trolley through a supermarket. Tending sourdough starters and baking bread, fermenting tea to make kombucha, turning fruit into jams, marmalades and chutneys – these are all time-consuming activities being taken up with enthusiasm. There is a sense that food is worth it. Going to the time and trouble of preparing food for others has long been entwined with deep-held ideas of nurturing and hospitality. Cooking a meal for others is an expression of affection – carried out by mothers and fathers for their children, by romantic partners for each other, by people for their friends.

Time is at once fluid yet implacable. Despite much human ingenuity being expended, we find it impossible to genuinely reproduce its effects on shaping and creating our food. We need to recognize what a truly remarkable ingredient time is – find pleasure in its proper use in the whole food chain, from growing and making to cooking – in short, to make time for it.

SECONDS

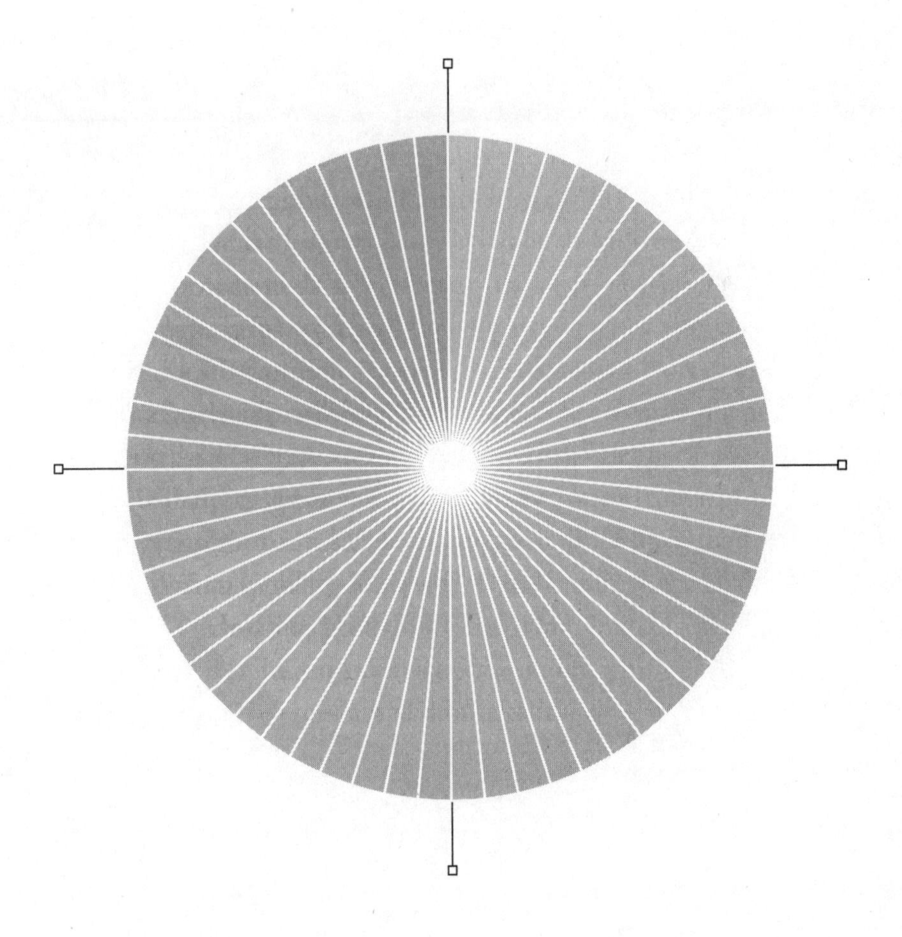

SECONDS

SECONDS

These tiny slivers of time move rapidly and inexorably. Their passage is so fast that very few foods are cooked in a matter of seconds. This means that despite the relentless social pressure to speed up our cooking – to provide instant gratification – the Seconds section of this book remains aptly short. But split-second timing – requiring an eagle-eyed vigilance and focus – can be called for in the kitchen to prevent such cook's calamities as chocolate seizing, custard curdling, mayonnaise splitting or simply overcooking. That knowledge of when a dish is ready – the second at which to stop – comes with experience. When it comes to consuming food, there is also an immediacy to our ability to taste. The prompt way in which our body responds to what we eat is a biological defence mechanism that helps prevent us from unknowingly swallowing bitter foods that might be poisonous. The sensual way that chocolate begins to melt within a few seconds of contact with the human body was the starting point for exploring this fascinating foodstuff.

THE IMMEDIACY OF TASTE

There is a vivid immediacy to how we taste. Place a piece of food in your mouth and you will begin to taste it within 1 or 2 seconds. We take this phenomenon for granted, yet within that fraction of time a lot has happened. The human mouth and throat contain on average around 10,000 taste buds; the ones on the tongue are found on small structures known as papillae. Within these taste buds are taste receptors, which contain microscopic, highly sensitive taste hairs (microvilli). When food is placed in the mouth, soluble chemicals within the food known as tastants dissolve on contact with saliva and these are detected by the taste hairs, triggering the taste buds to send taste messages via nerves to the brain. Our taste buds are able to detect five basic tastes: sweet, salty, sour, bitter and umami (savoury). These fundamental tastes detected by the mouth, however, form only part of what we perceive when we taste something. Our nose is hugely important in creating our overall taste 'pictures', which is why food tastes so dull when our ability to smell is impeded by, say, having a cold. Smell is vital to our perception of taste because when we eat, molecules from the food travel up through the back of our throat, reaching olfactory nerve endings in our nose and resulting in smell messages being sent to our brain. Smells received in this way – that is retronasally – are treated by the brain as being signals from the mouth. The information received this way forms what we 'taste'. Our olfactory sense is acute. Our noses are able to detect thousands of different smells, and it is this that allows us to recognize so many hundreds of flavours. Watch professional tasters at work and you will notice how they smell as well as taste; think of a wine expert swilling the wine around inside a glass to maximize the surface, aerating the wine and intensifying its scent, then sniffing deeply before drinking. This is done for a very good reason, as the wine's aroma will form such a large part of the wine's flavour profile.

The capacity to taste is something we take for granted, but it is central to our ability to enjoy food. We are able to discern a huge gamut of flavours, from the subtle tang of natural yoghurt or the juicy sweetness of a ripe peach to more assertive tastes, such as bitter coffee, fiery chillies or pungent garlic. As we age, however, our capacity to taste declines. Taste buds go through a two-week cycle of birth, death and renewal, but as our bodies become older, fewer taste buds are renewed. Furthermore, our sense of smell becomes less powerful, also affecting our facility for perceiving flavour. As our perception of flavours is muted, the rich, diverse world of food becomes very dull indeed. The next time you eat or drink something delicious, pause and reflect on what a life-enriching ability being able to taste is.

THE CRAFTING OF CHOCOLATE

Sitting at my desk, I reach over to a bar of chocolate – placed there for 'emergencies' – break off a square and place it in my mouth. Within just a few seconds what was solid at room temperature has begun to melt, with the warmth of my mouth triggering this process. Part of the alluring appeal of this extraordinary, much-loved food is that sensual melting on contact with our bodies.

Chocolate is created from the seeds of the cocoa tree, *Theobroma cacao*. Nowadays, chocolate confectionery is both readily available and affordable, but for most of its long history, cocoa was enjoyed in liquid rather than solid form. The origins of our relationship with cocoa can be traced back to Mesoamerica. Archaeological evidence shows that ancient civilizations, including the Olmecs and Mayans, used cocoa beans to make a drink. Cocoa, similarly, had a special place in Aztec society, a symbolic ingredient roasted and used to make a luxurious beverage flavoured with spices, consumed by the Aztec elite. Cocoa's discovery by Europeans came about through the Spanish Conquistadors, the Aztec empire falling to Cortés and his army in 1519–21. Chocolate, the exotic and novel beverage made from cocoa, credited with many properties including aphrodisiac ones, made its way into Spanish courtly circles and from there into European society. To make the drink, cocoa beans were ground into a paste, sold in cake form, which was then diluted with water or milk to form a thick, rich drink. By 1657 chocolate was on sale in London, advertised as 'an excellent West Indian drink called chocolate'. Samuel Pepys, a man always with an eye for the novel and the fashionable, writes several times of drinking chocolate.

Eating chocolate, as opposed to drinking chocolate, came to us by way of the Industrial Revolution, when the invention of ingenious machinery allowed cocoa beans to be processed in new ways. A number of today's major chocolate companies played an important role in developing chocolate as we know it. In 1847 Francis Fry, of the British Quaker cocoa company Fry & Son, mixed together cocoa powder, sugar and cocoa butter to form a bar of chocolate – a seminal moment. In 1879 a Swiss chocolate manufacturer called Rodolphe Lindt invented a machine called a conche, which slowly ground and mixed the chocolate mixture, reducing the particle size to create a smooth, rather than granular, texture. Lindt also increased the amount of cocoa butter added to the chocolate, thereby increasing the smoothness of the texture.

These inventions allowed eating chocolate to be mass-produced and affordable. Chocolate's useful ability to be melted and re-formed saw the rise of chocolatiers in countries such as France and Belgium. Chocolatiers need to first melt and temper the chocolate. Tempering is a process that involves taking the chocolate to a sequence of precise temperatures, which depend on the type of chocolate used. This realigns the crystal structure in chocolate, giving it a more stable form. When correctly done, tempering creates gloss and snap and allows the chocolate to be moulded and turned out successfully, so it can be shaped into bars or used to make confectionery, such as truffles. The skill of chocolatiers was appreciated and the results of their labours valued as luxurious treats.

Today chocolate manufacturing around the world is dominated by a handful of large companies, such as Mars Inc, Mondelez International and Nestlé S.A. The 1990s, however, saw the rise of a craft chocolate movement in both America and Europe, focused on exploring the potential of cocoa varieties and different countries of origin. In recent years there has been a growth in the number of what are often called 'bean to bar' craft chocolate–makers, usually working on a small scale, who source their cocoa directly from the growers. 'There was a coming together of good-quality cocoa beans becoming available, chocolate-making machinery becoming far more affordable and people such as chefs and tech enthusiasts becoming fascinated by the potential to make great chocolate,' explains Spencer Hyman of Cocoa Runners, a company which showcases and sells craft chocolate bars sourced from around the world. 'When we set up Cocoa Runners in 2013 there were three people in the UK crafting chocolate from beans they were directly sourcing – we now know of around forty. In America there were around twenty or thirty; there are now over 300.' Chocolate contains over 400 distinct flavour compounds, with flavours ranging from fruity or floral to earthy or spicy. The chocolate bar is now being used by committed chocolate-makers as a way of expressing cocoa's character. 'Normally with chocolate, because of the way it's been mass-produced, while you might get some nice flavour upfront, that will be it. What you get from a little square of craft chocolate is an almost wine-like complexity of flavours developing and evolving,' asserts Spencer.

One of my favourite producers is Danish chef turned chocolate-maker Mikkel Friis-Holm, a tall, genial bear of a man, who talks with great intelligence, knowledge and humour. Open-minded and intellectually curious, Mikkel has been pushing the boundaries since 2007. The care that he takes in making his chocolate begins with the meticulous sourcing of the cocoa beans he uses. 'I spend a lot of time on relationships with the growers. I've been to Nicaragua, Honduras, the Dominican Republic, to meet the farmers and the projects, to talk and set down protocols for fermentation.' This willingness to engage with those processing cocoa beans on the ground bore fruit when he asked what would happen if during the five-day fermentation (during which cocoa develops its characteristic 'chocolate' flavours) the beans were turned three times, rather than twice, as was habitual. In 2011, using Chuno beans from two separate batches, fermented in these two different ways, he produced two bars: Chuno 70% Double Turned Single Bean and Chuno 70% Triple Turned Single Bean. Taste them side by side and the differences are apparent: while both are richly flavourful, the Triple Turned is the more mellow of the two, with the Double Turned noticeably brighter, with more acidity and a zingy, tongue-tingling finish. 'The triple turn had more oxygen going into the fermentation pile, which increased the heat earlier in the fermentation and so you would have had the temperature of 124°F for 12 hours longer,' he explains. Experimenting with chocolate-making in this way and being able to compare the results had never been done before: 'Chocolate geeks over the world were astonished,' grins Mikkel. This fascination with the potential to make chocolate differently – his adventurous challenging of preconceptions – makes him a true trailblazer, one whose work is followed with real interest by others in the world of fine chocolate. 'I think chocolate is the most complicated product of them all, with so many flavour compounds. There are all these possibilities, but no one's exploring them,' he says emphatically. 'What's hyper-interesting about working with cocoa and chocolate is to be in an area where I can actually influence it.'

Over the winter of 2014–15 he set up a new Friis-Holm chocolate factory, located in peaceful, rural surroundings just outside Copenhagen. Chocolate-making is in progress when I visit, and the air is richly scented with chocolate. Watching Mikkel and his team hard at work brings home to me the amount of labour and the number of stages involved in transforming the huge sacks of cocoa beans into those sophisticated chocolate bars. First the cocoa beans are roasted, then cracked and time-consumingly winnowed, to remove the husk from the edible cocoa nibs inside. These nibs are ground into a thick paste with cocoa butter, the natural fat extracted from cocoa beans. This chocolate mixture is then conched. Peering into the large machine (12 feet long and 8 feet wide), I see and smell smooth, deep brown liquid chocolate, being moved by a granite roller beneath the surface in a fluid wave across the machine – a 'Charlie and the

Chocolate Factory' moment. The conched chocolate is next tempered in a tempering machine. The liquid chocolate is deposited into chocolate-bar moulds, with the glossy bars of warm chocolate – so shiny that their surface reflects the tiles behind them – carefully placed on a cooling machine. Once cool, the bars are wrapped by being passed through a large, intricate machine that evokes a Heath Robinson illustration from the Professor Branestawm books of my childhood. This is an industrial process, but whereas a large chocolate-maker will take cocoa beans and turn them into chocolate bars in 8–12 hours, here at Friis-Holm that process from bean to bar takes around 4 days. To begin with, Mikkel roasts whole cocoa beans, rather than nibs, a process which takes around 30–50 minutes. Roasting cocoa nibs (pieces of cocoa bean) would be much quicker, but brings with it the risk of burning the smaller pieces, while roasting whole beans 'retains more flavour'.

Each piece of equipment has been painstakingly researched and selected, with the longitudinal conching machine especially important. It is a piece of vintage equipment, around 100 years old, consisting of a shell-shaped trough with a granite roller. 'They don't make them like this any more. Companies now do dry conching, which needs the aid of an emulsifier. Dry conching was introduced because it shortens production time.' Why go to the time and trouble of sourcing and refurbishing this conching machine, I ask. 'Because it makes better chocolate,' Mikkel says simply. 'Its plasticity is unrivalled. We do around 30–48 hours of conching for each batch of chocolate.' The lengthy conching process, he explains, affects not only the texture of his chocolate, making it very smooth, but also, crucially, its taste. 'The warm water in the water jacket helps some of the polyphenols and volatiles evaporate, so by conching we remove excessive acidity and round off the flavours. When you hear that slap inside the machine, it's aerating the chocolate and in that way taking out unwanted acidity.' Taking the time to drive off the volatiles in this way enhances the flavours of his chocolate.

His award-winning range of chocolate bars, packaged in matte paper in pared-down Scandinavian style, has been created with such clarity as to give, in effect, a chocolate education. Rather than simply offering chocolate from countries of origin – something which Mikkel sees as merely a starting point for exploring cocoa – Friis-Holm bars feature chocolate made from cocoa blends, such as La Dalia, sourced from Nicaragua, and from single varieties of cocoa beans: Indio Rojo, from Honduras, and Nicaliso and Johe, both from Nicaragua, among them. He also experiments with levels of cocoa content, offering, for example, two Dark Milk bars, both milk chocolate bars made with higher than traditional cocoa content, each made from the same Nicaraguan cocoa, but one with 55%, the other 65%, allowing for a direct comparison. What is striking about Friis-Holm chocolate bars is the depth and variety of cacao flavours encountered. Some offer fruit notes – ranging from bright citrus to red berries – while others are spicy or earthy. These flavours are delivered in luxuriously smooth form due to the generous amounts of cocoa butter used in their production. Sampling these bars is a rich, sensuous experience. Mikkel has not forgotten that good chocolate is a treat.

Given Mikkel's creative energy, when I visit his factory, I am not surprised to learn that he is engaged in a new chocolate project. On this occasion, he is experimenting with ageing his chocolate bars. 'I think it's interesting to talk about chocolate as a live thing that has development. All chocolate-makers know that their chocolate changes in flavour a month or 2 months after making. If you tasted the same chocolate freshly made, then again after leaving it for a month, you would taste two different chocolates. There is a magic to the bonding of the molecules with time, but I think very young chocolate straight off the line could be as interesting as one where it's levelled out.' The starting point for this enterprise came about when Mikkel first came across a batch of Rugoso cocoa beans and, despite industry advice to the contrary – on the grounds that its flavour was too aggressive – made some chocolate with it. 'When I tasted the freshly finished chocolate it was – wow' – he pauses expressively and grimaces to convey the shock – 'very powerful, lots of tannins, made your teeth squeak! But I thought, this is exactly the same sensation I get when I drink a young red wine with lots of tannins. Usually those tannins will transform and make more intense flavours, and that is exactly what happened after a year and a half with this chocolate! The very fresh unripe plum-flavour notes had matured to ripe, almost prune notes, like a port. The spiciness, which had been like an angry bee that wants to get out, was still there, but it had mellowed. It was the best chocolate I had on the shelf,' he chuckles happily. 'What I would like to do is put this chocolate out in the market at three or four different ages, offer 1 ounce bars at 6-month intervals. In chocolate, no one has ever done that before.'

UMAMI: SAVOURY DELICIOUSNESS

The desire to produce food that we will enjoy eating, food that tastes good, has long been part of cooking, and this involves an understanding of how taste and flavour work. If one looks at a contemporary taste map of a human tongue, one finds sweet, salt, bitter, sour – and umami. The fifth taste – often described as 'savoury', 'meaty' or 'brothy' – is subtle in comparison to the four primary, easily identifiable tastes, and its discovery is a comparatively recent one in the history of taste. Eat a small piece of good Parmesan cheese and stop to focus on experiencing its long finish – a satisfying, salty-sweet richness; this is an example of umami. It is often described as having a 'mouth-filling' texture, creating the sensation of weight and depth. Other foods that offer an umami hit are Marmite and soy sauce. Umami is noted for its ability to enhance both saltiness and sweetness. In his foreword to *Umami: The Fifth Taste*, food writer Harold McGee observes that it remains 'somewhat mysterious – it's a more complex sensation than the other tastes and seems to depend on them to be fully appreciated'. Umami was identified and named by Professor Kikunae Ikeda, an academic at the Department of Chemistry of Tokyo Imperial University in the early years of the twentieth century. A soup prepared by his wife using *kombu*, the dried seaweed which, together with dried bonito flakes, is used to make *dashi*, the tasty, traditional Japanese stock, was the starting point for his investigation. Eating the soup one lunchtime, he realized that in the flavour of the soup there was something that was missing from the traditional map of four tastes and set out to research what this might be. In 1909 Ikeda announced in the *Journal of the Chemical Society of Tokyo* his discovery of a fifth taste: umami, 'savoury deliciousness'. He linked the presence of umami to glutamate (an amino acid and neurotransmitter); glutamic acid, naturally present in many foodstuffs, becoming glutamate when cooked or fermented. Further research took place following this discovery, and Japan's Umami Information

Centre now describes umami as 'the taste of glutamate, inosinate and guanylate'. Scientific research has also confirmed the presence of umami receptors on the tongue. Recent years have seen interest in umami spread both beyond laboratories and outside Japan, with acclaimed chefs around the world exploring how to use it in their kitchens. Among the results have been dishes such as Heston Blumenthal's scallop, birch syrup, bergamot and coral royale, David Kinch's Love Apple Farms tomatoes with Parmesan, and Virgilio Martínez's charred octopus, purple corn and seaweed – all created to offer the diner a culinary experience rich in umami.

It is noticeable that many of the foods high in glutamate, and, therefore, umami, are foods that require time to make. It is associated with processes such as drying – in the case of *kombu* (dried kelp seaweed), dried porcini or morel mushrooms, and sun-dried tomatoes – and fermentation, as with Thai fish sauce. Parmesan cheese, often cited as a Western ingredient high in umami, is an example of a cheese that is aged for several months. Hanging meat for several days notably increases its savoury properties. Ingeniously, however, Professor Ikeda, having isolated the chemical that causes the taste sensation umami, set to work stabilizing it. By mixing glutamate with salt and water, he created and patented monosodium glutamate, known as MSG, a white crystal. Named *aji-no-moto* ('essence of taste') by Ikeda, this is usually sold finely ground as a handy flavour enhancer. MSG's ability to enhance food by simply adding a pinch of it saw it embraced with enthusiasm by domestic households in Japan and China, as well as by food manufacturers around the world. It is an ingredient in Japan's Kewpie mayonnaise, a condiment that enjoys a cult following. In the West, MSG now has a poor reputation. During the 1960s it became associated with excessive thirst and headaches, symptoms known as 'Chinese Restaurant Syndrome'. Since then, however, the scientific community has cast doubt on the idea of Chinese Restaurant Syndrome and its link to MSG. Critics of MSG, however, see it as an unnecessary additive, a speedy, chemical short-cut substitute for a taste that traditionally took time and skill to create in foods such as cured meats, blue cheeses or rich stocks. While MSG is viewed with suspicion, the idea of umami is much in vogue in Western cuisine. Its proponents point out that the presence of umami allows for less salt to be used in dishes, while achieving satisfyingly flavourful results. The quest for umami has seen the rise of easy-to-use, umami-rich flavour enhancers made from natural foods. Ingenious food producers now offer umami pastes made from umami-rich ingredients such as tomatoes, garlic, miso and matcha, or umami powders made from shiitake mushrooms, salt and seaweed. In the meantime, *aji-no-moto,* or MSG – offering the convenience of tastiness in powder form – continues to be widely used in food manufacturing and fast food around the world.

FIFTY SHADES OF CARAMEL

Sugar and heat are all that is required in order to make caramel. As a child, I remember the fun of heating sugar and water until it caramelized, pouring out the golden syrup onto a tray, waiting for it to set and harden, and cracking it into shard-like pieces that I enjoyed as sweets. On the face of it, therefore, caramel seems simple. It is an ingredient that has many uses in the world of pâtisserie and confectionery. In restaurant kitchens, caramel is classically combined with egg custard to create French desserts such as crème brûlée and crème caramel. As the sugar is heated, the caramel changes colour in stages and the different caramels created have a range of uses. Pale caramel is used to glaze petits fours, while darker caramel, with its more intense taste, is used to flavour desserts. The visual potential of caramel – the ability to draw out the syrup into fine threads that harden as it cools – is epitomized in that French pâtisserie classic the croquembouche, for which legendary French chef Antonin Carême gave an early recipe in his 1815 cookbook, *Le Pâtissier Royal Parisien*. This elaborate creation is a spectacular pyramid of choux pastry balls, bound together and decorated with threads of caramel.

Recent years have seen the rise in popularity of salted caramel. My favourite version of it is made by British chocolatier Paul A. Young, whose award-winning salted caramel truffles – glossy domes of chocolate filled with smooth, rich salty-sweet caramel – are one of his perennial best-sellers. A visit to renowned pastry chef Claire Clark's workplace kitchens, housed on a light industrial estate in London, offered a glimpse into the intricate realities of working with caramel professionally. 'Caramel is really important to pastry chefs,' Clark tells me, 'up there with chocolate and nuts. It's used for a lot of things, not just sauces: a finishing item, a flavour and a colourant.'

Belying its sweet taste and the way in which it is used to create dainty treats, there is a ferocity to the making of caramel. Caramelizing sugar is dangerous because of the high temperatures required to trigger caramelization and the tenacious way that melted sugar retains heat. When you learn to make caramel as an apprentice chef you are never left unsupervised, because of the temperatures – and so the risk of getting serious burns – and the fact that the process goes so rapidly. The quickest way of making caramel, Clark demonstrates, is what she calls 'direct caramel', using solely sugar. First, a heavy-based pan is preheated over a high heat, then Clark adds just enough white sugar to coat the base of the pan. As it hits the hot metal it 'smokes' and begins to colour at once. More sugar is added in stages, Clark stirring in each addition thoroughly. The sugar has to be added gradually to prevent burning at the base, creating lumps of hard sugar. Within a few minutes she has created a dark caramel, the colour of chestnuts; perfect for a butterscotch sauce, sticky toffee pudding, nut brittle . . .

There is also another method Clark uses to make caramel: the 'indirect' way, in which sugar is combined with water before being caramelized. Creating caramel this way takes a few minutes longer, because of the time taken to drive off the water. The advantage to making caramel this way is that the process is longer, offering a more nuanced control over the colour of the caramel. The indirect method, or 'wet' method, brings with it, however, a higher risk of crystallization, when the melted sugar starts to form crystals once again, creating flaws and streaks in the caramel that can't be corrected. Clark takes certain measures to prevent this: using good-quality sugar and adding a touch of glucose syrup to the pan with the sugar. Water is poured in, enough to cover the white sugar, but not a huge amount. 'You don't want too much water,' she comments, 'as the longer it boils, the more risk of crystallization.' Having stirred it to 'slush', she puts the pan over a fierce heat, in order to bring it quickly to the boil and so minimize the risk of crystallization. Soon, the surface of the syrup is covered with a layer of bubbles. 'Don't agitate the pan at this stage,' she stresses. As the sugar and water mixture boils away in a hypnotic fashion, I realize that Clark, although relaxed and chatty in manner, is keeping a vigilant, hawk-like eye on proceedings. As she talks to me, she frequently brushes the sides of the pan down with a wet pastry brush: 'I don't want sugar crystals falling back in and I don't want splashes to colour on the sides, then fall back into the syrup.' Glancing at the mixture, she notes, 'It's bubbling ferociously and the bubbles are very small, so I know that at this stage it's not even at soft ball.' The 'soft ball' stage is reached by sugar syrups at 235–240°F; at this stage a small amount of syrup dropped into cold water forms a flexible ball. A few seconds later, however, she points out that the bubbles are slowing and getting bigger, becoming visibly denser: 'It will begin to caramelize soon,' she forecasts. Sure enough, the syrup around the edge of the pan is taking on a golden hue. Clark pulls the mixture off direct heat, lets it rest, then gently but thoroughly shakes the colour through the rest, spooning out a little to set on a tray. Returned to the heat, the mixture speedily turns several shades darker. 'This is what I would say is almost the perfect colour, a golden colour. The only way to stop the process now is to plunge the pan into cold water to cause the caramel to seize.' Despite being taken off direct heat, I can see the implacable way in which the caramel changes colour over a matter of seconds, going from a dark brown ('good for a crème caramel, as the bitterness would contrast with the custard') to a very dark brown. Part of Clark's skill is being able to 'predict' the changing colour of the caramel as time passes. The samples on the tray have hardened and set, taking on a clear, glass-like appearance, ranging in colour from a delicate yellow to a treacly black-brown. Each of these stages, Clark explains, differs not only in colour, but in flavour, with the darkest colour being the most bitter. 'This very dark caramel we would use for colouring; for example, we use it to colour a caramel cheesecake that we make. The bitterness works well with the sweetness of the mixture.'

As a pastry chef, Clark will choose indirect or direct caramelization depending on what she wants to use it for. In making Florentines, for example, she would use the indirect caramel method, adding double cream and butter to sugar and honey and caramelizing it but only cooking it to a pale golden colour. She knows that the mixture, thickened with chopped nuts, peel and candied fruits, will be shaped and baked, with the time in the oven deepening the hue from pale gold to brown and causing the caramel to set.

One of the big issues for professional pâtissiers is achieving consistency of colour. The ability to work speedily with caramel once it's at the desired point is vital. For example, making a nougatine – a mixture of light caramel and nuts – requires the caramel nut mixture to be rolled out speedily while it's at the right temperature. If it cools and becomes hard, it can be 'rescued' by being softened in the oven and worked again, but the oven's heat will affect the colour. 'Making caramel is very, very visual. Repeating a colour consistently is a massive challenge,' Clark tells me. 'At the French Laundry [the California restaurant where Clark was head pastry chef], it was unacceptable to have anything but the exact shade of colour; there was no leeway! A thermometer won't help, because there are so many factors. It's about how fast you cool it down, how fast you work. It is very reliant on the skill and experience of individuals. To work quickly with caramel takes time to learn.' In the world of pâtisserie, the sheer speed of caramelization means that it requires skill and patience to harness and use it effectively.

THE BREVITY OF BLANCHING

I'm standing in my cousins' sleek, orderly, spacious kitchen in their apartment in Kuala Lumpur, Malaysia. Angela, my cousin's Singaporean wife, is about to cook one of Singapore's much-loved national dishes: Hainanese chicken rice. Angela is a wonderful, knowledgeable cook and I am always keen to watch her at work in her kitchen. A large saucepan of water, flavoured with a few slices of ginger and spring onions, is at a steady boil on top of the stove. Angela, dextrously using an oven mitt and tongs, slowly and gently manoeuvres a whole chicken into the water, ladling hot water over the chicken as she lowers it in. Once the bird has been totally covered by the water, she removes it. 'You can see the skin has started to tighten and contract,' she points out. She returns the water to the boil, carefully immerses the chicken in it one more time, then removes it once again. She then returns the chicken to the pan to be gently poached in the water until just cooked. I have witnessed blanching in action, the word derived from the Old French *blanchir*, meaning 'to whiten'. The reason for this preliminary stage when cooking the chicken, Angela explains, is to help achieve the desired texture. The whole cooking process, beginning with the carefully controlled introduction of the chicken to hot water, is designed to ensure moist, soft chicken. 'With chicken rice, what's particularly prized is the silkiness of the flesh – it shouldn't be hard or dry.' The same cooking technique, she tells me, is sometimes used in Chinese cooking when making a pork stock. Pork is added to boiling water, then quickly removed. In this case, however, the water, clouded with proteins released by the pork, is discarded. The pork is then slowly simmered in fresh water to create stock. A clear stock, free of murky protein, is the desired goal.

Blanching – immersing an ingredient in boiling water for a short period of time, measured sometimes in seconds, sometimes in minutes – is one of the speedier culinary tasks performed by cooks. Anyone who has enjoyed a plate of tender, yet not soggy, leafy green vegetables such as choy sum or bok choi with oyster sauce in a Chinese restaurant will have experienced for themselves how effective this rapid cooking technique is with leafy vegetables. 'For me, I put the vegetables into boiling water and as soon as I see the bubbles coming up, so it's about to boil, but hasn't returned to the boil, I take them out,' Angela observes. In cases where timings are fine, the blanched food is then immediately plunged into iced water to halt the cooking process at once, 'refreshed'.

Blanching in cooking is often used as a preliminary stage; for example, softening fresh vine leaves or cabbage leaves so that they can be wrapped easily around a filling. As a method it has a range of uses. It is used to help remove recalcitrant, hard-to-peel skins from, say, tomatoes, peaches or almonds, the contact with boiling water so brief that it does not affect flavour or texture. It is a way of removing powerful smells from ingredients such as garlic, onions and bacon, muting their potency, and to remove bitterness from certain vegetables. In the world of frozen foods, blanching is important in freezing vegetables, as it stops enzyme reactions within the vegetables – that freezing doesn't halt – which would adversely affect colour, flavour and texture. It is a brief, versatile cooking method. As the scholarly *Oxford Companion to Food* deftly observes, 'Blanching lasts for a shorter or longer time according to what is being blanched and for what purpose; but it never lasts long.'

THAT COFFEE RUSH

It is 8 a.m. on a summer morning in the heart of Florence. Swifts screech overhead and scooters whiz past outside on the grey flagstoned road. I'm standing in Robiglio, a bar discreetly located in a side-street in the heart of the city, just a short walk from the imposing edifice of the Duomo. Immaculately clean, with its marble floor, dark wood counters and mirrored walls – not to mention the tempting display of sweet-scented pastries – Robiglio is classic in appearance, an institution, well loved by locals. As I wait and watch, a steady stream of people comes in through the bar's open doors: men in suits clutching briefcases, smartly dressed women, a cook in chef's whites from the restaurant next door, students, burly workmen. Each customer arriving is clocked by the staff, acknowledged with a 'Buon giorno.' The barman, in his black-and-white uniform, asks the same question over and over again: 'Caffè?' – the word rising interrogatively on the second syllable. The answer is always 'Sì.' Everybody, without exception, orders a coffee. Regular customers are recognized, their coffee prepared instantly. As each order is given, the barman deftly uses the gleaming, metal espresso machine, emptying out old coffee grounds with firm taps, adding fresh coffee to the basket, tamping it, locking the brew handle in place, positioning small cups to receive the coffee, frothing milk, placing small, thick, white china saucers in front of each customer at the bar in preparation for their drink. As he works, he wipes the nozzles with a cloth, keeping them clean; there is a sense of pride about his work, a feeling of professionalism. Customers stand at the bar drinking the hot, bitter drink. Service is prompt and the coffee is drunk rapidly. Pastries chosen from the display are simply wrapped with a square of paper and passed across to customers, to be held in one hand, eaten while standing or pacing the floor. What is noticeable is the rapidity of the entire process, with just a few minutes passing between a customer entering the bar, consuming a coffee and leaving it. The Italian bar in the early hours of the morning, I realize, offers brief encounters that are at once purposeful and pleasurable. The scene taking place before me in Florence is re-enacted in bars in villages, towns and cities across Italy. Long before the shops wind up their shutters the hardworking bars of Italy are open for business, serving the essential fuel with which Italians begin their day.

For centuries, coffee has been a drink associated with stimulation. In coffee mythology, coffee was first discovered in Ethiopia by a goatherd named Kaldi, who noticed how lively his goats became after eating the fruit of the coffee tree. The seeds of the berries of the *Coffea* plant contain caffeine, an alkaloid compound which acts on our bodies, stimulating our central nervous system. The beverage coffee, made from the roasted seeds of the *Coffea* plant, is the world's main source of caffeine. As we drink a coffee, the caffeine is rapidly absorbed and spreads through our bodies. Within 20 minutes its effects begin to be felt, apparent in an increased alertness and ability to concentrate. This comes about because caffeine blocks adenosine receptor sites. The accumulation of adenosine in our brain tells us we are feeling tired, so its blockage by caffeine makes us feel more 'awake'. After around 3 hours these effects fade, causing a coffee 'crash'.

Coffee as a drink made its way from Africa to the Middle East via Arab traders, arriving in Turkey, and then in Europe, with 1645 seeing the opening of a coffee house in Venice. London's first was opened in 1652 by the enterprising Pasqua Rosee, in an alley by St Michael's churchyard, off Cornhill. A handbill entitled *The Vertue of the Coffee Drink*, published that year to spread the word about Rosee's venture, highlighted one of this exotic new drink's properties: 'It will prevent Drowsiness and make one fit for business.'

Over time, espresso has become *the* emblematic coffee drink: sophisticated, dark, intense, an adult drink, designed, as its name suggests, to be consumed at speed. It came about as inventors searched for faster ways to produce cups of coffee for customers in busy cafés. During the nineteenth and twentieth centuries the power of steam was explored and brought to the making of coffee, though the story of how the espresso machine was created is – as with many scientific inventions – drawn-out and complex. It started in 1884, when Angelo Moriondo was granted a patent for a machine which used steam and water to create a coffee beverage. In 1906 Bezzera and Pavoni launched their own coffee machine, the Ideale, at the Milan Fair. The Ideale used steam pressure to push water and steam through ground coffee, creating what Pavoni termed 'caffé espresso', coining the term. While it had the novelty of being made quickly, the coffee extracted at high temperature and low pressure was both bitter and weak. The major breakthrough came about in 1938, when Achille Gaggia patented his invention, an espresso machine which used a piston mechanism to force hot water through the coffee at high pressure. This created an espresso characterized by its *crema*, the name given to the distinctive, creamy-looking head of foam formed on the surface through the release of carbon dioxide from the coffee during the pressurized process. The power of the machine meant that a shot of flavourful coffee could be made in around 30 seconds. In 1948 Gaggia founded his eponymous company to produce his machines and the Gaggia Classic machine became a familiar sight in Europe's coffee bars.

Recently a speciality coffee movement has grown in the UK, with the influence of Antipodeans credited as a major factor in both creating the demand for quality coffee and supplying it. The opening of the Antipodean-style coffee shop Flat White in London in 2005 was an important moment in this story. While per capita coffee consumption in the UK has, in fact, stayed largely static, through the popularity of chain stores such as Starbucks and Costa Coffee, coffee is nowadays a visible presence on UK high streets. Interestingly, within the UK market, sales of instant coffee have been declining in recent years. Just one in five coffee drinkers are drinking instant coffee more than once a day, with this figure falling to 8% for people in their early twenties. The search for convenience and speed when it comes to making coffee, however, has seen the rise in popularity of coffee-pod machines, pioneered by Nestlé with Nespresso. These machines use foil-sealed pods of pre-ground coffee which can be simply placed inside the machine, where they are pierced and water is forced through, resulting in a coffee a mere 12 seconds later. Such is their growing popularity that coffee pods entered the basket of goods used by the Office for National Statistics to measure inflation in 2016.

James Hoffmann, World Barista Champion 2007, has been a key player in creating and shaping the UK's speciality coffee scene. In 2008 Hoffmann, together with Annette Moldvaer, World Cupping Champion 2007, founded Square Mile Coffee Roasters, a roasting company based in London's East End, wholesaling their own-roasted beans. When I see that a coffee shop or café is using Square Mile coffee, it is a sign that this is a place that takes coffee seriously and my chances of finding a good cup here are high. Hoffmann, moreover, is a man focused on creating a thriving coffee culture in Britain, sharing his expertise through his writing and a range of Square Mile events. Arriving at the roaster, I pass through an unassuming door in an industrial estate into a coffee-scented, coffee-centred world. A tall, slender, youthful figure, Hoffmann talks with great articulacy and a dry wit, as precise in his speech as he is in his coffee-making.

It is the roasting stage, when heat is applied to them, that transforms these green beans, sourced from coffee-growing regions – Africa, Asia and the Americas – into the familiar dark brown beans. For craft roasters, the skill and attention given to roasting is an important element that allows them to create 'their' coffees. This is what I wanted to see: Square Mile's large roasting room – assailed by the noise of machines, whirring fans and blaring background music – where I watch a batch of green coffee beans from Ethiopia being roasted for the company's espresso blend. With a note of pride in his voice, Hoffmann explains that the solid drum roaster I'm looking at is an old one built in the 1950s, bought and rebuilt by Square Mile around 4 years ago. 'Speciality roasters somewhat fetishize vintage equipment,' he notes, 'but its appeal is often due to the cast-iron drums themselves. Modern roasters don't use the same kind of cast iron. It's like an old skillet; there's something lovely about cooking with those compared to new ones. The precision and the robustness with which this was made is amazing to me.' This drum roaster holds around 66–77lb of beans at a time: 'In the great scheme of the global coffee industry, this is laughably small. In the speciality coffee industry, this is medium-sized.' Inside the drum, he explains, are paddles that mix and move the beans. 'Much of the work during roasting is done by air. So you generally have a very large gas burner that heats the drum and also the air that flows through the drum. A big fan pulls the air through the drum from the burner, then expels it.'

With an ear-battering clatter, 72.6lb of green coffee beans are blown up the tube to sit in a hopper above the drum. The temperature inside the drum is key to successful roasting, and with the refit of the roaster came the addition of two accurate temperature probes, one measuring the temperature of the beans and the other measuring the temperature of the air, allowing the person roasting to track the progress of the roast against the desired roast profile on a screen. 'These are our target outlines, what we know tastes good. The function of time and temperature is incredibly important.'

Once the air inside the drum has reached the right temperature, known as the 'charge temperature', around 390°F, the coffee roaster opens the hopper, sliding the green beans into the drum in a great rush, with their arrival cooling down the hot air dramatically. 'You can roast very, very quickly, get the whole job done in 90 seconds, but this is really done just with instant coffee, which doesn't taste that great,' James explains. 'Some semi-commercial roasters might roast in 6–8 minutes. We tend to roast slower. A slow roast is generally preferred by speciality companies, as we find it gives a much more developed, sweeter, complex product. For some companies, slow roasting might go all the way to 22 minutes. For us, that's a bit too slow; it's too dark and it's spent too long in there. For us, for espresso, in this machine, 14–16 minutes would be the window we work in. If we were roasting a filter coffee in this machine, then more like 10–12 minutes, depending on batch size.' Ever alert to nuance, however, Hoffmann goes on to explain that it is not just the total roasting time which is an issue: 'It is the progression of temperature, how long the coffee takes to pass through different stages, how slowly or quickly the beans reach certain temperatures. It's time *and* temperature.'

The first stage in roasting is known as drying, requiring a huge amount of energy to evaporate the water from the beans: 'We can't get any browning reactions yet because of the presence of water, as it limits the temperature. At the moment, the beans smell quite plant-y, quite green.' At Square Mile, they take time at this stage to get out as much water from the coffee as possible, and to remove it evenly. 'Slowness helps us to do that evenly. If we went very quickly, the outside of the beans would lose their water and begin browning, before the beans inside could, so you'd end up with an uneven roast with some undercooked, which is not desirable.' Driving out the moisture so thoroughly is important, explains Hoffmann, because otherwise the coffee will tend to retain grassy flavours and sour notes. 'There are lots of ways the roaster can make small mistakes which cause the roast to be relatively uneven. We find the window of "delicious" quite narrow.'

Once the drying has been completed, the beans begin 'yellowing'; a tryer is used to draw out a small sample of beans to look at them and check progress. After an expert assessing glance, Hoffmann says enigmatically, 'We're about 4-ish minutes from cracking at a guess.' Waiting for the coffee to progress through the roasting process requires, I realize, a curious combination of patience and alertness. 'The thing about roasting,' says Hoffmann with feeling, 'is that it's this strange, strange mixture of boredom and stress!' Sure enough, in 4 minutes' time, alongside the relentless noise of the beans whirling in the drum, I hear a faint crackling sound within the roaster, a noise that reminds me of popcorn popping in a covered saucepan. This is what is known as 'first crack', created by gases, largely carbon dioxide, and water vapour being ejected with force from inside the beans, cracking them open as it pushes out. Hoffmann shows me a sample bean, and I can see that it does indeed now have a tiny split, the crack from where the carbon dioxide has left. Around 2 minutes later the beans are released with a great clattering noise into the cooling tray, a satisfying climax to the process. The freshly roasted coffee beans are around 390°F, and the aim is to cool them down as quickly as possible using a large fan, stirring them to ensure even cooling. Within just 3 minutes the coffee reaches close to room temperature. 'The coffee would taste just a little less sweet if it cooled too slowly,' Hoffmann explains.

The coffee I've seen roasted so carefully will be used to create an espresso coffee blend, the floral flavours of the Ethiopian coffee combined with two other coffees to make Square Mile's distinctive blend. An old-school coffee roaster would pre-blend the beans for an espresso blend, then roast them in one go. Different beans, however, are different sizes and densities, so Square Mile prefers to roast them separately, then blend them together. Furthermore, the decision to take the roasting for espresso to first crack, but not on to the audible stage known as 'second crack' when the beans reach around 435°F, is characteristic of contemporary speciality coffee roasting. 'For a long time, roasters roasted to second crack for a richer, heartier style of espresso,' Hoffmann tells me. 'The modern wave of coffee is more interested in clarity and sweetness than bitterness, so we tend to roast a bit lighter. The downside is that it is very easy to roast too light and then you can get a sour, harsh product. All these things are a very fine line . . .'

Once cooled, the freshly roasted coffee beans are packed in order to preserve their quality, then date- and batch-stamped, with samples of each batch set aside for quality-control purposes. Although Square Mile ships out their coffee on the day of roasting or the following day, Hoffmann recommends resting the coffee for a few days before using it. This is in order to let the carbon dioxide created by roasting dissipate, as its presence affects how the coffee is extracted. The cafés that Square Mile supplies with coffee wait at least 7 days before using a batch of freshly roasted espresso. Domestic users, however, often open the coffee at 3 days, since it will take them longer to use up a bag of beans and they want to use them while they are at their best, which is within a month of roasting. Storing the coffee in an airtight container, kept dark and cool (though not the fridge), helps keep it at its best during this period.

Having watched the roasting process, Hoffmann and I move to the quieter surroundings of the training room to discuss how coffee the drink is made. Broadly speaking, Hoffmann explains, there are two types of brewed coffee: infusion – 'coffee and water just sit and hang out together' – and percolation, where the water passes through the coffee. An Aeropress combines both, as the coffee is first steeped and then percolated. Coffee the beverage is created through water's contact time with ground coffee, and the type of grind affects the brewing time. Generally, percolation coffees use fine-ground coffee and infusion coffees a slightly coarser grind. Without heat as an accelerator of the process of extracting coffee, the time taken to create a coffee stretches, so a cold-brew coffee made using medium-to-coarse coffee can take 24 hours.

Whether making coffee in a café or at home, using infusion or percolation methods, starting with freshly ground coffee beans is key to making good coffee, in Hoffmann's opinion. 'Pre-ground coffee is disappointing and a pale comparison to fresh ground, because three things happen: you lose those aromas that are so wonderful; oxygen begins to turn the fats rancid; and new, less desirable flavours begin to appear.' For this reason, and on principle, Square Mile only sells whole coffee beans, not ground coffee. Also 'super-important' when making coffee is the weight of coffee and the weight of water. The number of factors and the way in which they impact on the results are obviously a source of fascination for him: 'It's not a simple, straightforward process,' says Hoffmann, 'that's how we can hold all those coffee-brewing competitions!'

As we sit and talk, Hoffmann courteously asks if I'd like a coffee and I request a filter coffee. I watch with interest as, working deftly but precisely, he weighs out the required amount of beans, grinds them and makes the coffee. At every stage, from the way he adds a little water to the ground coffee at the start to moisten it and allow it to swell, a practice known as 'the bloom', to the careful weighing of the water, there is a meticulous quality to what is being done. After 2½ minutes' brewing I am presented with a cup of black filter coffee. There is, pointedly, no milk or sugar on offer. I sip it cautiously: the flavour is noticeably sweet but with an acidity there too, balancing it and giving it a refreshing quality. It is a world away from the thin, unpleasantly bitter filter coffee one finds sitting on hotplates in conference centres. This is, I realize, the most enjoyable cup of filter coffee I've ever drunk.

The espresso, Hoffmann explains, is at the heart of the contemporary craft coffee movement. 'Here's the thing – the way speciality coffee tends to infiltrate a market is through espresso, because espresso is quick to make; a good espresso might take

25–30 seconds to make, whereas a filter coffee is more like 2–3 minutes to brew. Using an espresso machine, you get a very small, concentrated essence of coffee which turns out to dilute very well. So, espresso is great, because not only can I make it in 25–30 seconds, I can then turn that espresso into a dozen different things – latte, flat white . . . I can buy two bits of equipment – a grinder and an espresso machine – and make a wide range of coffees.'

Having seen the care and thoughtful attention given at every stage of the process, from the roasting of the coffee beans to the making of coffee at Square Mile, I ask Hoffmann to talk me through what lies behind a disappointing cup of coffee. He begins with the coffee used to make it: green beans bought on price, rather than quality, which are then roasted very quickly in order to get more throughput and more coffee and quenched to cool them, so putting water in the beans. This low-grade coffee is then pre-ground before selling, meaning that its flavour and aroma diminishes. A long supply chain ensures that it is 3 months before the coffee is used, although it remains well within its 18 months use-by date. The cup of coffee is made quickly and carelessly by someone using a dirty machine. 'This will deliver you a bitter, caffeinated but not hugely enjoyable drink that demands milk and sugar. There's a very good reason why most people want milk and sugar in their coffee; it's because most coffee is terrible.'

MINUTES

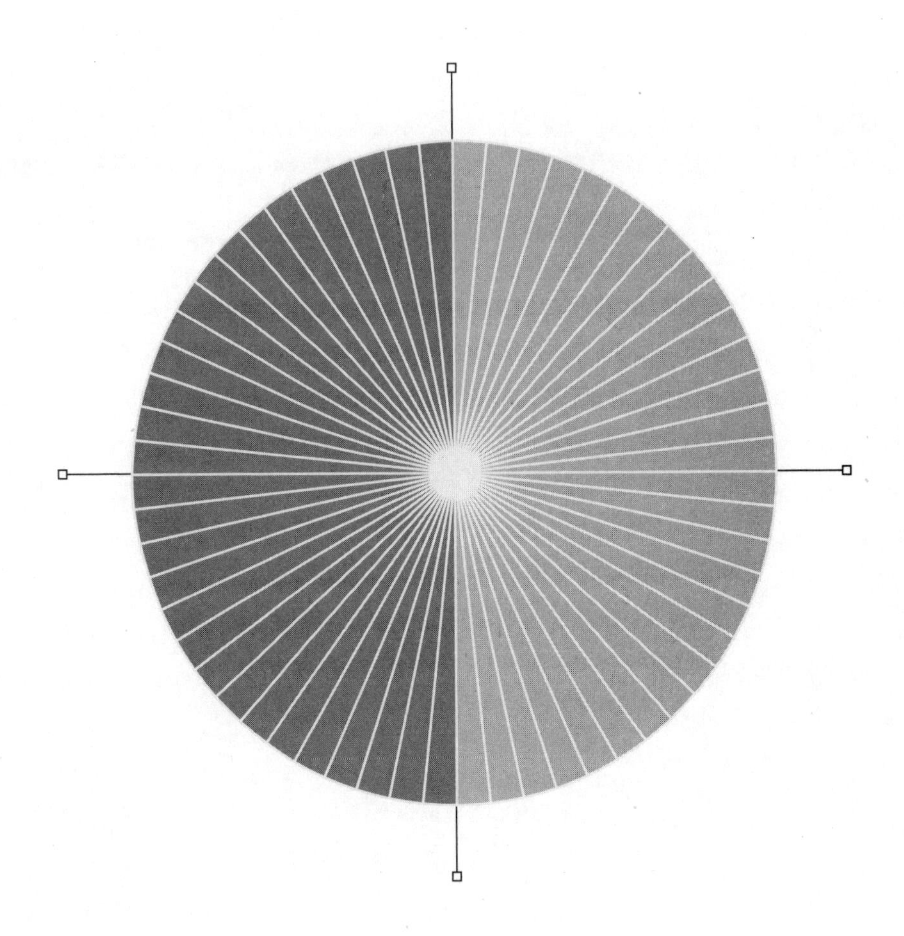

MINUTES

MINUTES

Minutes are the building blocks of time spent in the kitchen. So many of the foods we eat can be cooked in minutes, while tea and coffee – those much-loved beverages – are made in minutes. They are admirably flexible amounts of time, small enough to be used to measure culinary processes that happen quickly. And – with 60 of them in an hour – they can also be used to measure slower ones, such as cooking rice, browning meat or making mayonnaise. The versatility of minutes in the world of food is apparent from the range of subjects covered here. It is noticeable that in our impatient world, where speed is valued, using as few minutes as possible to make food has become a goal we rush towards. Microwaves, of course, cook food in minutes, fast food is served in minutes, and the demand for cookbooks measuring the cooking of meals in minutes continues unabated.

FRESH

'Freshly clicked', reads the slogan on the side of a supermarket van driving past me in the street. Freshness in food – eating something soon after it has been harvested, killed or prepared – has long been valued in many cultures. To begin with, of course, during the many centuries before refrigeration, freshness was linked to the fundamental fact of foods such as meat or fish being safe enough to eat. The human fascination with freshness, however, is also linked to the pleasure we take in eating. Food that is fresh is valued for the quality of the culinary experience it offers. People place a real premium on freshness.

But food chains now are often extraordinarily long and complex. Everyday ingredients are sent on long journeys, even within their country of origin. Consequently, for the ordinary shopper in the Western world, access to what seems to be such a simple thing – fresh food – is hard to find. Supermarkets are adept at the trappings of freshness: displays of fresh fruits and vegetables, pots of fresh green herbs, the scent of freshly baked bread. But much of the produce will have been imported from far away, the herbs are hydroponically grown, lacking the flavour of outdoor-grown herbs, the bread is industrially produced, arriving ready-formed and simply baked in an oven at the store, without the flavour and texture of artisan bread. The popularity of farmers' markets in America and Britain is, in part, because they offer a different model of food shopping, where one can buy food direct from the producer. The freshness of vegetables on offer is often cited by customers as a reason to shop at these markets. Freshness has not totally vanished, but it requires localism and short food chains. Industrialization and globalization have led to long supply chains – often international – which, by their nature, cannot offer true freshness.

Real freshness requires a knowledgeable clientele that understands ingredients. Around the world – in Latin America, Africa, the Middle East, Asia, parts of Europe – the habit of shopping daily for food in a market is taken for granted. On a recent trip to Singapore, I visited Tiong Bahru 'wet market' early one weekday morning. Arriving at 8 a.m., I passed a steady stream of local shoppers, people leaving the market, carrying plastic bags of fresh fruit, vegetables, fish and meat, their purchasing already done in the cool of the morning before heading off to work. This market, and many like it internationally, continues to thrive because of the commitment by its customers to shopping for the freshest ingredients every single day.

In Britain, a country whose food retailing is dominated by the supermarket model, the habit of daily shopping was replaced during the 1960s and the 1970s by the convenient notion of the 'weekly' shop. We still, however, hold on to the idea of freshness being especially important for certain foodstuffs. Dairy used to be one of these. From my childhood, I remember the daily milk van, trundling slowly on its rounds in London's streets, delivering pints of fresh milk to household doorsteps early each morning. Today the milk round has largely vanished from Britain's streets, with milk a cheap supermarket loss leader, bought in bulk once a week.

Similarly, with fish. To eat fish and seafood as soon after catching it as possible is not only practical – because it deteriorates so quickly, so offering risks to health – but gloriously pleasurable. Seafood restaurants around the globe take advantage of their access to fresh fish and seafood. Think of tangy ceviche in Lima, simple grilled fish by the edge of the Bosphorus in Istanbul, Marseille's iconic dish of bouillabaisse, created from the catch of the day, a sushi breakfast in Tokyo's famous Tsukiji fish market, tasty grilled sardines enjoyed by holiday-makers on Portugal's Algarve coast . . . In Chinese gastronomy, live seafood restaurants are popular, offering tanks of crustaceans such as lobsters and crabs and fish as part of the 'menu'. At these, diners select what they want to eat, whereupon the chosen seafood is fished out, despatched and cooked at once. But the loss of independent fishmongers on high streets means that the chance to enjoy very fresh fish is increasingly rare. Tellingly, whereas in 1992 Britain's National Federation of Fishmongers had 2,000 members, today its membership has dwindled to around 400. One fishmonger who has found an appreciative clientele for what he has to offer is Rex Goldsmith, whose small shop – the Chelsea Fishmonger – is housed in a 100-year-old fishmonger's tucked away on Chelsea Green, a short stroll from the busy King's Road in London. A tall, affable man with a ready grin, Rex has been up since 3.30 that morning when I meet him at his shop at half seven. His first task of the day is to drive from his home in Surrey to Billingsgate Market in London's Docklands, where he does some top-up buying, to fill in any gaps in his stock. The majority of what he sells, however, is delivered each morning from Cornwall by a company in Newlyn that he's bought from for 27 years. Most of the fish he sells is caught off the coast of Cornwall or Devon by small dayboats, which return to harbour with their catch. The stock bought by Rex over the phone is then delivered early next day. While I watch, Rex and his assistant work busily unloading large white polystyrene boxes – which squeak loudly as they're moved – from the van. As they open them, I spot huge turbot, sea bass, mackerel, live lobster. There are signs of freshness to look out for, Rex explains: bright eyes, bright red gills, a sheen to the skin; also, through touch, firmness and stiffness, the latter meaning rigor mortis is still set in. I notice, too, that though I am surrounded by fish, there is very little 'fishy' smell, the lack of odour another sign of how fresh the stock is.

In this exclusive part of London, serving a knowledgeable, discriminating, cosmopolitan customer base, it is vital for Rex's business that he offer top-notch, spanking-fresh fish and seafood. His business, he tells me, is built on relationships, working with trusted suppliers who sell him the best and his ability to buy well. 'I've got such respect for the fishermen,' he says, 'it's dangerous work.' As children walk by on their way to school and commuters head to work, Rex and his assistant are busy carefully arranging his purchases on crushed ice on the slab. While he works, customers walking by exchange greetings and banter. Realizing that one elderly, distinguished-looking gentleman wants to buy something, Rex pauses in his work to serve him: 'That was a very nice mackerel you gave me yesterday,' his customer remarks with satisfaction. The display of stock being assembled so carefully on the surface is, I realize, in effect Rex's billboard advertising his business. 'Flash', Rex calls his display, the show he puts on in his shop each morning it is open. Long, gleaming white cod fillets, scallops with bright orange roes, dark pink wild salmon fillets, dappled grey, brown, green flat fish, iridescent blue-green mackerel . . . As I look at it, I think that one instinctively recognizes true freshness when one sees it. Rex's skill, which he enjoys exercising, is in astutely judging quantities so that he can satisfy his customers' demands that day, without carrying over old stock. 'This will all go by late morning,' he says proudly, gesturing to the display.

Fresh produce, too, is much championed in the world of

restaurants. The 'farm to table', 'locavore' movement has seen chefs in many countries – such as Alice Waters at Chez Panisse in Berkeley, California; Alain Passard of L'Arpège in Paris; or Simon Rogan of L'Enclume in Cumbria, UK – working closely with farmers and growers and also setting up their own farms and gardens to supply them each day with flavourful vegetables, fruits and herbs. The second incarnation of influential chef René Redzepi's Noma restaurant in Copenhagen features an 'urban farm'.

Freshness in fruit and vegetables remains dear to us in the West, something tapped into by veg box schemes, delivering vegetables and fruit from the growers to our homes. Many people, given a chance, enjoy growing their own – whether, like me, it's simply a collection of potted herbs by the back door or on a windowsill, or planting vegetables and fruit in the garden. The importance of people having access to land in order to grow fruit and vegetables was recognized in industrialized Britain with the creation during the twentieth century of the allotment system, allowing people in towns and cities to rent small plots of land for personal productive growing. Such is the desire of people to grow their own that demand for allotments is high. The National Allotments Association estimates that over 90,000 gardeners are currently waiting for allotments. While I wait at the counter to buy meat, my local, genial butcher, Steve Noon – always up for a chat – has often talked to me of his allotment and the satisfaction of eating his own-grown produce. A July day visit to Noon's plot at Brook Farm Allotments in Whetstone, north London, reminds me how appealing well-tended allotments are. Sloping down towards the railway tracks of the Northern Line tube, the land, with its mown grass paths and neat plots, is a visible expression of the care and trouble taken by the allotment holders here. It is an immaculate, verdant, productive oasis. With palpable pride, Noon shows me round his lot, which is as neat and orderly as his butcher's counter. As we walk, we pause now and then to pick and nibble fruit: two varieties of gooseberries – one tart, the other a much sweeter dessert fruit – and an intense blackcurrant. 'Sadly, you've missed the cherries,' Steve observes. 'Wonderful crop this year.' Steve took on his plot 3 years ago, but explains that the first year was not a productive one, as the ground had to be dug over and the weeds removed. 'These were just shapes on the ground,' he tells me, pointing to the beds. Now, however, Steve and his family are relishing the fruits of his weekly labour here, from the potatoes, beans and peas to the fruit, transformed by his wife into jams to last through the year. 'If you can pick it and take it home and cook it within half an hour, it's a completely different flavour,' he tells me. 'Courgettes, corn on the cob, beetroot – anything you buy in the shops, no matter how fresh they say it is, it will taste different. Even potatoes, when you dig them up and cook them, well, the flavour . . . I'm no expert, but you can really taste the difference.'

LEARNING TO SAVOUR

The idea of savouring food – taking the time to taste it carefully and thoughtfully – can seem a thing of the past, one where all of life was a more leisurely affair. This slow, appreciative approach to eating was advocated by French gastronome Jean-Anthelme Brillat-Savarin, who observed that the 'careless, hasty eater' misses out on pleasures that a more considered approach brings. The experience of eating good food can indeed be enhanced by pausing to consider and appreciate its flavours as one consumes it. A flavourful length of finish is a hallmark of much carefully made food, and one that deserves noting by the person eating it.

I was reminded of the pleasures offered by savouring food at a day course offering a Level 1 Certificate in Chocolate Tasting, a qualification which sounded eminently pleasurable to attain. The course, in King's Cross, London, has been recently set up by chocolate expert Martin Christy, a co-founder of both the International Chocolate Awards and the International Institute of Chocolate and Cacao Tasting, and the fact that such a course exists is an interesting development in the fine-chocolate scene.

The first thing that strikes me is the international range of attendees; here are people from Belgium, Canada, Ecuador, Ireland, Korea, Spain and the United States. Interest in learning about chocolate and how to taste it, it would seem, is widespread around the globe. A tall man, with a scholarly, thoughtful air to him, Christy proves to be an effective and engagingly humorous communicator. The aim of the day, he explains, is to inform us about chocolate and to teach us to understand and appreciate it. Within minutes of his introduction, Christy declares, 'One of the biggest issues is that we are conditioned to eat chocolate very quickly. So the first thing is to slow down – we will do this so much that you're not going to properly eat chocolate until after lunch!'

Aromas form an essential part of flavour, and we are first trained to recognize key chocolate flavours by sniffing pots containing among other things tobacco, cinnamon, wood, raspberry jam. Next we each smell a spoonful of melted chocolate – a tantalizing experience as we are strictly forbidden to taste what's in front of us – analysing and discussing the scent. Christy similarly takes us through the five tongue tastes: sweet, salt, sour, bitter and umami, using tastings to reinforce recognition. Following a talk on the chocolate market and fine chocolate, we finally taste melted chocolate. Again, the experience is controlled. We eat it first while holding our nose, then, once it has been swallowed, release our noses. The rush of flavour inside my mouth when I release my nose is extraordinary. Christy surveys our surprised faces with satisfaction. 'For a millisecond you're transported to a cacao farm,' he comments. Length of flavour, he explains, is one of the most important indicators with fine chocolate; our brains are still processing the flavour archetypes within the small piece we ate. The noticeable length of the finish is a sign of quality. In order to create fine chocolate, such as that we tasted, good cacao is key, with the current level of interest in sourcing and using it to make chocolate a recent phenomenon in chocolate's long history. The need to support cacao farmers and value their work is a recurrent theme during the day.

The next sensory exercise brings home just how rewarding savouring a piece of fine chocolate can be. We are each given a piece of chocolate and told to snap it in half. As precisely instructed by Christy, I take one of the chocolate halves, hold it at arm's length, look at it, bring it slowly towards me, smelling it before I place it inside my mouth. We have been told sternly to let it melt inside our mouths, with no chewing or biting, which proves to be a slow, sensuous, pleasurable experience as the flavours of the chocolate unfold. Going around the room, the flavours identified by the group include citrus notes, passion fruit, fragrant honey, with a gentle tannic feel in the mouth as part of the aftertaste. 'Take your second piece,' says Christy. 'Now we're going to do chocolate speed eating. I'm going to count to three and then you're going to eat it as fast as you can, chew it in 7 seconds.' We rapidly do as he says. This time the flavours identified are noticeably limited: sourness, tannins and vanilla, with very little aftertaste save an astringent dryness in the mouth.

What happened with the rapid chewing, Christy explains, was that the cocoa butter – that fat that is solid at room temperature but which melts in the mouth, playing a key part in our sensual experience of chocolate – simply didn't have time to melt in the mouth. Cocoa butter delivers the flavour, so in allowing the chocolate to melt, we get the full complexity of flavour. The fat of the cocoa butter balances the tannins. Eating the same chocolate quickly not only loses the full benefit of the flavours, but also leads to that unpleasant dry aftertaste. So even good chocolate eaten quickly seems as though it's not very good. 'The moral of the story is melt, don't munch.'

A course in appreciating fine chocolate sounds truly indulgent, but as he talks, I realize that there is a distinctly political aspect. 'The idea is to create an army of consumers around the world to spread the message about good chocolate,' he tells me. Part of appreciating fine chocolate, he feels, is taking the time to eat it. Educating people that we should treat fine chocolate in the same way as a fine wine is for Christy a way of building up an appreciative audience who are prepared to pay for quality. 'If we want to have a chocolate bar and support the farmers, have fine cacao that the farmers actually get paid a decent amount for, then we have to develop a market strong enough for fine cacao. We need to create a market that is more like coffee, where speciality coffee is valued, where there is an understanding of what a good chocolate bar should cost.'

THE EXACTITUDE OF EGGS

Despite being an everyday ingredient, when it comes to cooking them, eggs invariably demand an eye to the clock. Such is the precision of time required to cook them, that eggs, uniquely among ingredients, have their very own timers in the kitchen: these include traditional, miniature, hour-glass-shaped timers or (for boiled eggs) heat-sensitive plastic 'eggs' which, when added to water, mimic the cooking process, changing colour as they heat up. So complex is the cooking of eggs that much has been written on the subject, with entire cookbooks devoted to this versatile food. 'If you want to learn how to cook, start with eggs,' advises cookery writer Delia Smith, known and loved for the reliability of her recipes, at the start of her book *How to Cook: Book One.*

One of the factors which make eggs complex to cook is that an egg is made up of two substances that behave very differently: the yolk and the white. The yellow yolk is rich and nutritive, while the white is predominantly made up of water, containing some proteins. In both the yolk and the white, the proteins are dispersed in water. As we cook eggs, the heating process causes the egg protein chains to unfold, then bond with each other, transforming raw eggs from a slimy-textured, slippery liquid to a solid state. When cooked, the yolk and the white begin to solidify at different temperatures, hence the possibility of having a soft-boiled egg with a cooked white but a runny yolk. The nuances of egg cooking are apparent in the variety of egg-centred diner lingo, with expressions not only for generic egg dishes – with 'wrecked', for example, meaning 'scrambled' – but for the specifics of how they should be cooked. The visually expressive 'sunny side up' means an egg fried only on one side with the white set and the yolk still soft, while 'eggs over easy' means eggs that are fried until set, then flipped over very briefly to just set the exterior of the yolk, while leaving it runny inside.

In several classic egg dishes, it is vital not to overcook the eggs, with cooking taking place in minutes, sometimes seconds. Many people prize the delicate texture of scrambled eggs or want their Spanish tortilla to arrive soft and oozy in the middle, rather than 'overcooked' to dryness. Even something as apparently simple as a hard-boiled egg can be overcooked, resulting in an egg that is tough, rubbery and sulphurous. On the other hand, undercooking eggs can provoke squeamishness: a soft-boiled egg where the white, as well as the yolk, is uncooked and liquid rather than set is very off-putting to many people. It is important to use the exact amount of cooking time in order to achieve the results you want.

One of the first things to bear in mind when it comes to cooking with eggs is their freshness – the amount of time that has passed since they were laid – since this affects their texture and what dishes they can best be used for. EU legislation requires that the maximum 'best-before' date on an egg is 28 days from it being laid, so the eggs one buys can be weeks rather than days old. In the absence of use-by dates, one easy way to test the freshness of eggs is to place them in a bowl of water. An eggshell is porous, so once an egg has been laid, moisture from the egg is lost through evaporation, the air pocket inside the egg becomes larger and the egg, in effect, becomes lighter in weight. When placed in the water, fresh eggs rest horizontally at the bottom, less fresh eggs begin to tilt up, while a very stale egg will float on the surface.

Certain ways of cooking eggs require fresh eggs, poached eggs being the prime example. 'The one and only essential condition in this case,' declares French chef and writer Auguste Escoffier authoritatively, 'is the use of perfectly fresh eggs.' A truly fresh egg when poached holds its shape well and has a smooth, creamy quality to its firm textured white, as well as a delicate freshness of flavour, which makes it a pleasure to eat. As an egg ages, it changes its texture: the white becomes runnier, the ovomucin (a protein within the egg white which holds it together) breaking down. The yolk absorbs water from the egg white, becoming thinner in the process and its membrane stretching and weakening. Try to poach an old egg and it will fail to hold its shape compactly, instead dispersing in the water. On the other hand, using older eggs, with their weaker whites, is often recommended for making meringues, as they can be whisked more easily.

Fresh eggs are also seen as key to the omelette, an iconic egg dish emblematic of the ability to cook well and confidently. 'Successful omelette-making is a skill which only comes with experience,' declared the renowned French chef Paul Bocuse. 'Practice is more important than advice.' Elizabeth David, in her famous essay 'An Omelette and a Glass of Wine', evoked its pleasures with characteristic elegance: 'As to the omelette itself, it seems to me to be a confection which demands the most straightforward approach. What one wants is the taste of the fresh eggs and the fresh butter and, visually, a soft bright golden roll, plump and spilling out a little at the edges.' In order to achieve a classic omelette, the consensus among food writers is that speed is of the essence. In *Mastering the Art of French Cooking*, by Simone Beck, Louise Bertholle and Julia Child, the recipe for l'omelette roulée specifies 'Less than 30 seconds of cooking'. The great food writer Richard Olney, an expert on French cuisine, concurs: 'Its execution demands less than a minute's time from the breaking of the eggs to the presentation at table.' In *French Provincial Cooking*, Elizabeth David avoids giving specific timings in her omelette recipe, describing a visual cue instead: 'When a little of the unset part remains on the surface the omelette is done.' A decent omelette pan, butter, eggs, the right level of heat and split-second timing are the fundamental elements to a good omelette. Despite the simplicity of the dish, the preoccupation with making it well continues to this very day. Type 'How to make a perfect omelette' into YouTube and the results total just shy of 450,000. 'As everyone knows,' observed David wryly, 'there is only one infallible recipe for the perfect omelette: your own.'

While speed and freshness are key to creating a good omelette, at the opposite end of the culinary spectrum there is a tradition of preserving eggs, notably China's historic delicacy, 1,000-year-old eggs (*pidan*), also called 100-year-old eggs or century eggs. The practice of making these goes back generations and involves coating the eggs in a highly alkaline mixture, often made from wood ash and lime, and setting them aside for months. During this period, the egg protein denatures, changing the egg's texture and colour, the yolk becoming greenish-black and the white a deep brown, while the egg acquires a pungent ammonia smell and a deeply savoury flavour.

It is noticeable that many great chefs enjoy the challenges that cooking an egg to perfection presents. French chef Alain Passard, of 3-Michelin-starred L'Arpège in Paris, is famed for his hot-cold soft-boiled egg – a dish so good that chef David Kinch of Manresa puts a version of it on his menu in homage. To make it, Passard cuts off the top of a fresh egg, pours out the white and, wittily, cooks the seasoned yolk in its natural eggshell container, floating it on the surface of simmering water like a small, bobbing boat. The freshly cooked yolk is served topped with cold crème fraîche and a touch of maple syrup, offering an intriguing contrast in temperatures. And playing with the cooking time of eggs is a hallmark of such ingenious chefs as David Chang, of Momofuku fame. Inspired by the Japanese tradition of cooking eggs in hot springs, Chang's Noodle Bar restaurants in New York and Toronto offer slow-poached eggs, cooked in their shells for 40–45 minutes in simmering water at between 140° and 145°F. The key to substantially longer cooking times is using lower than traditional cooking temperatures, with the sous-vide cooker – nowadays a staple piece of kit in professional kitchens – as this allows for long, gentle cooking at a specific temperature. At Benu, his sophisticated 3-Michelin-starred restaurant in San Francisco, Korean-born chef Corey Lee offers his own 'whimsical' take on China's historic *pidan*, using quail eggs (rather than the traditional duck eggs), which are brine-soaked for 12 days, then dried and aged for 4 weeks at 68–77°F. The traditional ammonia flavour is toned down in his version, Lee explains in his cookbook *Benu*: 'It's a beginner's *pidan*, so to speak: half a preserved quail egg served with a pickled ginger cream and a rich potage made with pork belly and just a touch of chilli.' Eggs – such a commonplace, affordable ingredient – continue to enthral.

A FISHY ON A DISHY

Across a range of cuisines – Chinese, French, Japanese, Indian, Italian, to name but a few – and assorted cooking methods – deep-frying, baking, steaming, grilling, poaching – cooking times for fish are measured in minutes. In his classic recipe for *truites au bleu* (blue trout), where trout are first freshly killed, then cooked at once in a court-bouillon, the French chef Paul Bocuse recommends: 'Allow 7–8 minutes for fish weighing about 150g (6 oz).' In my much-thumbed copy of *Yan-Kit's Classic Chinese Cookbook*, the recipe for clear-steamed sea bass advises steaming a fish of 700g–1.2kg (1½–2½lb) over a high heat 'for about 8 minutes until the fish is cooked'. Why does fish cook in minutes?

Well, the reason fish cooks so fast is because its flesh is delicate in structure. Fish flesh is fragile because fish live in water, which is denser than air and offers support. This means that fish don't need the strong interconnective tissues or heavy skeletons that land creatures require to support their bodies. Their shorter, finer muscle fibres make fish flesh noticeably tender and easy to eat. Furthermore, fish have far less collagen, the structural protein found in animal connective tissue. Meat collagen requires long cooking to break it down. This is why it's all too easy to overcook fish. We don't appreciate how quickly it cooks, with the results often sadly dry. During an interview, French chef Pierre Koffmann, discussing his fondness for eating fish, remembered a visit to a market in Sicily. 'Wonderful, fresh seafood,' he said, expressively, in his rich French accent, 'beautiful tuna, tiny red mullet . . . There was a little restaurant there, so we had lunch.' He grimaced at the memory – 'It died twice, eh!' – his horror at the overcooking vivid. His disappointment that poor cooking had wasted a precious resource is telling. Precision and attentiveness are required in order to cook fish and seafood well. Restaurants can acquire a cult following for their way with fish, being known for an impeccable rendition of one fish dish, like veteran British institution Wiltons, on London's elegant Jermyn Street, noted for grilled dover sole, or barbecue restaurant Elkano in Spain's Basque country, renowned for its whole grilled turbot, lovingly cooked over charcoal.

Despite its island credentials, Britain – bafflingly – is not known as a nation enamoured of fish, other than in its favourite, battered, deep-fried 'fish and chips' form. Despite the variety found off our coasts, our consumption centres on only five main species: cod, haddock, tuna, salmon and prawns make up 60–70% of the seafood eaten in the UK. This contrasts with other countries, such as Spain, where fish is consumed with relish. A visit to a Spanish food market, such as the Mercado Central de Abastos, in the heart of Jérez, shows stall after stall piled high with fresh, noticeably affordable fish and seafood – hake, sardines, octopus, razor clams – with patiently queuing customers waiting to be served. As the restaurants in the streets nearby demonstrate, fish is central to Spanish cuisine, cooked with care and skill. One woman determined to transform the British relationship with fish is the redoubtable, down-to-earth C. J. Jackson, CEO of the Billingsgate Seafood Training School. Housed above London's historic fish market, nowadays located in Canary Wharf in Docklands, the school offers training to the seafood industry and also teaches domestic cooks. Many of the courses here begin in the early hours, with a 6 a.m. start to catch the market before it closes, students then being shown how to handle and cook the fish and seafood they've glimpsed for sale.

An excellent, experienced teacher, Jackson is clear that fish and seafood requires quick cooking. 'Very often recipes tell you to cook it for far longer than it actually needs. It's a delicate protein. If you overcook you ruin it,' she says with characteristic forthrightness. During her classes, Jackson demonstrates different cooking techniques. 'You never know how bold someone is with, say, a frying pan. If you've got a hot enough pan, squid is done after 15 seconds on either side,' she observes. Her students are often genuinely surprised at how quickly fish cooks. She talks me through cooking lemon sole fillets, fried in butter in a hot pan. 'What I do is put four fillets in, skin side uppermost. I put one in, two in, three in, four in. Go and pick up a palette knife and in the time I've picked it up, they're ready to turn over. So you turn the first, the second, the third, the fourth. A couple of minutes later I lift them out and they're done. It's a really nice way of cooking lemon sole.' Cooking times, she points out, will vary not only with the size and thickness of the fish or fillet, but also with its texture, with dense-fleshed fish like brill or turbot requiring more time for the heat to penetrate the flesh than a delicate-fleshed fish like halibut. How to know when the fish is cooked through is also important, and Jackson outlines some useful indicators. With whole fish, it's when the eye turns white. For grilled fish, she suggests feeling the fish: 'You know it's cooked when the skin becomes loose and if you press it you can feel the flake of the fish underneath.' For a whole fish that has been slashed, the flesh visible through the skin will change from translucent to opaque when cooked. For unslashed whole fish, Jackson recommends 'running a knife along the lateral line of the fish, right down the middle, just to cut the skin, pull it back slightly, because you can see whether the flesh is still slightly glassy'. She sets the anxiety about cooking fish through in context: 'People are so terrified of undercooking fish, but, in actual fact, undercooking chicken or pork is far riskier.'

Jackson's motivation to teach people how to cook fish springs from a desire to open people up to the pleasures of eating it. 'My belief is that one of the reasons why in the UK we don't tend to like fish, or we have people who are a bit phobic about it, is because people were given overcooked fish as a child. When it's overcooked it becomes spongy and dry and a lot less appetizing. The thing is that fish is expensive and people are worried about ruining it when they cook it and, of course, most of the time it's ruined because they overcook it. It needs to be just cooked to perfection.' Her advice when it comes to cooking fish is firm: 'Be brave. Very, very rarely does it need another minute.'

THE MODERN MICROWAVE

The sounds it makes are all too familiar – a whir, a muted roar and the piercing ping that says, 'Get it now!' The microwave oven is a standard piece of home equipment in many countries in the developed world. Approximately 90% of households in America own a microwave, 92% in the UK and 86% in France. With the success of the microwave has come the rise of the microwaveable meal. Simply unpack it, remove the protective film layer, place it in the microwave and in a matter of minutes your meal is ready. Invisible rays have cooked your food for you – and fast! A ready-meal portion of spaghetti with meatballs takes just 2½ minutes to heat through in a microwave, compared to 20 minutes in a conventional oven. The time spent cooking the spaghetti and making the meatball sauce has been spent in a factory, allowing you simply to heat up the results to a temperature at which the meal can be safely eaten. This way of interacting with food – cooking and heating it with such speed – is a twentieth-century development, tracing its origins to work during World War II in the field of radar technology.

As with many scientific discoveries, there was an element

of chance to the discovery that microwaves could be used for cooking. During World War II, Raytheon Corporation of Waltham, Massachusetts, produced radar equipment for the military. While the war surged, one of Raytheon's employees, Percy Spencer, an expert in radar tube design, noticed that when he stood in front of a magnetron emitting microwaves (short electromagnetic waves), a chocolate bar in his pocket melted. Intrigued, Spencer investigated further, successfully popping popcorn using microwaves and, so legend goes, exploding an egg. He went on to build the first prototype microwave oven, a metal box that contained the rays, and in 1947 Raytheon constructed Radaranges, an early line of microwave ovens. In contrast to today's compact, kitchen-surface boxes, these were almost six-and-a-half feet tall and cost over 3,000 dollars. The 1960s, however, saw the development of compact microwave ovens, allowing America to become an early adopter during the 1970s.

Microwave ovens work by bombarding food with waves of electromagnetic radiation that move at the speed of light. Friction caused by water molecules within the food reacting to the electric field is what heats up the food. A revolving turntable ensures that this heating process takes place evenly. The reason why a microwave heats up food so rapidly is that the microwave radiation penetrates food to a depth of 1 inch. In contrast, the infrared energy used in conventional cooking works first on the surface and then passes from the exterior to the interior, which takes time. This speedy penetration of heat means that a baked potato cooked in the microwave will take merely 8–10 minutes, rather than the hour required by an oven. What that microwave-baked potato will lack, however, is crisp skin. It turns out that traditional microwaves fail to create the Maillard reaction, the browning process that gives so much flavour to foods from bread to meat. Ready-meal producers, therefore, carefully provide the desired texture already, before their products are warmed up.

Generations growing up with access to microwave ovens as the main piece of cooking equipment are becoming unaware of traditional, more time-consuming cooking skills. One of my son's student acquaintances at university was astonished to learn that scrambled eggs could be cooked on the stove. In the West, microwave ovens no longer have the allure of exciting new technology that they possessed during the 1960s and 1970s, having become instead a default, rather dull piece of kitchen kit. In our time-pressed world, though, they remain a perceived essential for many households.

TIME FOR TEA

Tea – made from the tea plant, *Camellia sinensis* – is the most widely consumed beverage in the world after water. Broadly speaking, its long history – which dates back over two millennia, with the first written reference to it recorded in Szechuan, China, in 59BC – follows a trajectory from an expensive luxury, enjoyed by aristocratic elites, to something that is now commonly available and affordable. Notably, though, for much of its history, tea has been a drink rich in association with ceremony and ritual – from its historic role in Buddhism and the development of the intricate Japanese tea ceremony, to the culture of Russian samovar-based hospitality and Korea's tea ceremony. There is a long tradition, across many cultures, of treating tea with respect. Ancient texts, such as the Chinese poet Lu Yu's *The Classic of Tea*, written between 758CE and 775CE, give an insight into the care expected during its preparation. On making tea, he advised that the water in the cauldron should be brought to the boil in three stages – 'fish eyes', 'pearls from a gushing spring along the rim' and 'galloping waves' – but no further, as it would then be 'boiled out and should not be used'. The same attention was applied to the drinking of tea, Lu Yu advising that 'when you drink it, sip only. Otherwise you will dissipate the taste.' Historically, both the preparation and the drinking of tea was a measured affair.

With among the highest per capita consumption of tea in the world, Britain has a special relationship with the beverage. The British love affair with tea has an aristocratic heritage, with Portugal's Catherine of Braganza, who married England's King Charles II, credited with introducing the custom of tea-drinking to the British court during the seventeenth century. Tea consumption in Britain increased, with the nineteenth century seeing London become the capital of the world tea trade. For centuries, China had the monopoly of growing and making tea. It was the drive by Britain, enabled by its position as an imperial power, to find an alternative source that resulted in the cultivation of *Camellia sinensis* var. *assamica* (Assam tea), indigenous to India. With the rise of plantations and the growth of the black tea industry, 'tea' in Britain became synonymous with 'black tea'. Britain's tea habit has a dark side, with the First Opium War (1839–42) against China triggered in part by the British desire to pay for the costly commodity by supplying the Chinese with opium in return, and the use of indentured labour in India. In contrast to this brutal realpolitik, however, the 1840s saw the beverage gain the distinction of having a meal and a time of day linked to it, with Anna, Duchess of Bedford, credited with popularizing the custom of afternoon tea. The stereotype of the tea-drinking Briton, pausing to have a cup of tea, is a pervasive one – think of cartoonists Goscinny and Uderzo's *Asterix in Britain*, with the British buoyed by the addition of 'magic' herbs (which turn out to be tea leaves) to boiling water to fight off the Romans.

It is ironic, therefore, that as a nation famously characterized by our love of tea, the making of it in Britain has become such a brief, perfunctory affair. A tea-bag is placed in a mug, a kettle is boiled with a roar of noise, boiling water is poured over the tea-bag, almost instantly the water takes on a dark brown colour and – a few seconds later – the tea-bag is taken out and discarded. The cup of tea is ready, a hot drink with a taste so familiar that you hardly notice its flavour. When it comes to drinking tea outside the home, things are little different. Order a cup of tea in most cafés – even cafés that pride themselves on the range and quality of their coffee – and within minutes you are handed a disposable cup of scalding hot water inside which, dangling on the end of a string like a small, drowned creature, is a tea-bag, to be removed with a wooden stick. With tea, a drink that in the past was required to be freshly made with leaf tea on demand, the tea-bag – quick to brew and easy to dispose of – now dominates its making. In 1969, in the United Kingdom, 97% of people made tea using leaf tea, with only 3% using tea-bags. Today, those numbers are reversed, 97% of tea being made using tea-bags.

Britain is not alone in this. Time for tea is being squeezed, so to speak, around the globe. Countries that previously took the time to use leaf tea are turning to the tea-bag. There are no global statistics for how much of the tea grown around the world is now used in tea-bags; however, experts comment on a rise in tea-bag consumption in countries where it wasn't traditional, including India, China and Africa. At global tea conferences, where major manufacturers present their products, there is a tremendous growth in products now offered in tea-bag form.

The first patent for a tea-bag was issued in 1896 to one A. V. Smith of London. But it is an American tea merchant, Thomas Sullivan of New York, who is credited with popularizing them in the early twentieth century by sending out samples of tea in small silk bags. His customers were meant to use the tea without the bags, but opted instead for the convenience of immersing the sachets of tea in hot water and so the tea-bag was invented.

Snip open a standard tea-bag, shake out its contents and you'll see a textured, granular powder made up of tiny fragments of tea – so small, indeed, that one would be hard put to recognize that these are indeed pieces of leaf from *Camellia sinensis*, the tea plant. The type of tea used in most tea-bags is what is known as CTC tea, standing for 'crush, tear and curl' or 'cut, tear and curl' tea. The curt term alone offers a revealing insight into the mechanization of tea-making that this process brought with it. The distinctive flavours in tea are created by oxidizing the tea leaf, and the invention of the CTC process sped up the making of black tea enormously. Developed in India by William McKercher in the 1930s, the CTC process was a significant step in the industrialization of tea production. It used a withering machine, rather than the traditional bamboo trays and withering racks, and replaced the rolling of tea leaves with a machine that effectively minced the leaves into very small particles, which then oxidized speedily in dryers. The CTC process produces a uniform tea, one which is strong and infuses quickly, but is notably lacking in aroma. Its commercial success is closely linked with the increased use of the tea-bag, with CTC tea perfectly suited to the form. The rise of the utilitarian tea-bag in turn plays a key part in tea's continuing mass consumption in cultures where convenience and speed of delivery is prized.

And yet while the tea-bag continues its seemingly implacable rise in usage, there is also a rise in leaf tea appreciation around the world. In continental Europe, Germany takes tea very seriously, with customers prepared to pay high prices for quality teas in their cafés. In Britain, while the black blended tea market (e.g. Tetley's and PG Tips) is flat, the market for speciality teas (an umbrella term applied to green, white and oolong, single estate and flavoured teas) is increasing at 7% per annum. In 2013, Unilever, the world's biggest tea company, acquired Australia's upmarket T2 chain in order to extend its portfolio in premium teas. North America – where over 80% of the tea consumed is in the form of iced tea – is also seeing a growing interest in speciality teas. DavidsTea, a Canadian specialist tea company founded in 2008, now has over 200 stores in Canada and the United States. It is noticeable how, in contrast to high-adrenalin coffee, the marketing of speciality tea pitches it as a means to relax, as a drink that offers a leisurely, civilized pause amid the hectic activity of modern life. It is a sign of the times that the tea-bag market is responding to this interest in leaf tea by offering better grades of tea, including sometimes whole leaf tea, in tea-bag form. Upmarket tea-bags are often larger, allowing for a better brew. Assorted tea-bag shapes, such as the tetrahedron or pyramid or the round bag, are now being used, but the difference in form is more presentational than practical.

Tea connoisseurship has a distinctly international flavour, with teas from around the world bought and sold globally to those interested in fine teas. Located in London's West End,

Postcard Teas, founded by Timothy d'Offay, has a cosmopolitan following, shipping out specialist teas to customers abroad. Such is d'Offay's fame as a tea expert that visitors to his shop come from many countries, notably from Japan, where his book on black tea has gained him a cult following. Customers for the teas he sources and blends include the food writer Nigel Slater and restaurants such as the River Café and Quo Vadis. In contrast to the hurly-burly rush of Oxford Street just around the corner, Postcard Teas is a serene establishment. Simply entering its aesthetic premises – a single, uncluttered room with its long wooden tasting table, distinctively packaged teas from countries including China, India, Japan, Sri Lanka and Taiwan on black shelves along one wall, and a few choice pieces of teaware on display – I always begin to feel less frenetic. During the late eighteenth century, in a satisfying piece of historic continuity, these premises at 9 Dering Street were home to a grocer called John Robinson, who would have sold tea. Calm, thoughtful and softly spoken in person, d'Offay, it soon becomes apparent, is a man on a quest to restore respect to the making of tea. For him, that begins with using leaf tea. As an importer and retailer of fine teas, he doesn't sell or deal in tea-bags. 'The main thing I have against tea-bags,' he tells me, 'is that you can't see the tea, you can't smell the tea through the paper and you taste the paper when you make it. If you use leaf tea, you see the tea, you smell the tea and as you make the tea, you start to learn about these little leaves, you start to have a relationship.'

In contrast to the mechanized mass-production of the CTC method, the teas carefully sourced and sold by d'Offay are very much made by human beings. They are made on small tea farms (each under 15 acres) that he has personally visited and knows. A highlight of his range is the Tea Master Teas, featuring teas such as China's Gold Medal Ba Xian (Eight Immortals), handmade by renowned tea-maker Master Lin from a 400-year-old tea tree, and Master Xu's Fragment of Imperial Robe. 'These people are artists, not artisans,' says d'Offay passionately. The range of teas available at Postcard Teas reflects the ways in which tea is made, the skill with which tea-makers process the leaf, transforming it into types of tea with very different aromas and flavours. In effect, the categories of tea – white, green, oolong, black – reflect different levels of oxidization. The tea leaves are 'plucked', then 'withered' to reduce the moisture and make the leaves more supple. The leaves are then fired or dried in the case of un-oxidized teas like white or green, or cut or rolled to break the cell walls and encourage oxidization, then fired to fix the tea at the required point for oolong and black teas. In countries such as China the tea-maker's ability to judge the timings of these different stages such as withering or firing is an important part of their expertise.

Since the discovery of tea, much has been written about how to make the 'perfect cup'. During the nineteenth century, polymath and tea enthusiast Francis Galton conducted numerous experiments in his quest for his holy grail – 'tea that was full bodied, full tasted, and in no way bitter or flat', devising mathematical formulae featuring the temperature of the water and the steeping time for the leaves. George Orwell, in his 1946 article 'A Nice Cup of Tea', declares that 'the best manner of making it is the subject of violent disputes', then gives his own eleven rules, 'every one of which I regard as golden'. For his part, however, d'Offay eschews hard-line generalizing on the subject. 'There's no one perfect formula for brewing tea. Brewing times vary for the different types of tea, partly because of the leaf and how the tea was made,' he observes. 'It's a flexible formula involving the volume of tea to water, the water temperature and the time taken.'

Tea-bags made using CTC tea brew so rapidly because the tiny particles of tea inside react very quickly with hot water, colouring and flavouring it within seconds of contact. Making tea from leaf tea, depending on the type of tea being used, does require minutes rather than seconds. As I realize from watching d'Offay brew a range of teas, the brewing time for leaf tea is a few minutes at the most. Certain teas, it becomes apparent, require more attention to detail than others. Yimu Oolong, a Taiwanese tea noted for its fragrance, is made by Mr Hsieh on his one-acre farm in Ming Jian in Nantou County, an area famous for its teas. The leaves are plucked, outdoor withered, then indoor withered for 12–24 hours, tossed in bamboo cylinders to break the cell walls and so cause oxidization to around 15–20%, with Mr Hsieh then firing the tea to fix it, rolling it and shaping until it reaches the correct stage of dryness. When making Yimu Oolong, d'Offay first washes the tea in hot water for 20 seconds, then pours off the water. The reason for this initial washing, he explains, is not to clean the leaves, as some people think, but to start to unfurl the leaf. 'When this tea was made, the leaves were rolled, so there's little contact between the leaf and the water. As they unfurl, there's more contact. Of course, you could pour hot water on and leave it to steep for longer, but if you simply do that you won't get as good a first infusion. This washing starts the unfurling and, therefore, the infusion will be a fuller one.' Next, he fills the kettle with fresh water and brings it to the boil. The just-boiled water is poured into a jug to slightly cool the water, bringing it down to around 200°F. This water, in turn, is poured over the washed tea leaves and the tea is left to brew for around 2½ minutes. Having tasted it to check that the tea has brewed as he wants it, it is then totally poured off the tea leaves. I am handed a small, delicate cup of oolong tea, yellow-gold in colour with a noticeable floral fragrance and a lingering sweetness in the mouth.

The care that d'Offay took to drain off the tea from the tea leaves is explained when he tells me that, by stopping the steeping process as soon as the tea is ready, these same tea leaves can be used again to make more infusions, in fact, up to four or five more. Had he simply left the tea leaves in contact with the water they would have continued to steep, resulting in bitter, 'stewed' tea. 'Tea was historically exceedingly expensive,' he points out, 'so that's partly why it was infused more than once.' In Europe people resold tea which had been steeped, then dried. With Chinese and Japanese tea culture, although tea was also a luxury, 'the infusions are to do with respect and connoisseurship, with using the tea to its full potential, enjoying the different states of it, the sweetness, the bitterness, the different characteristics. I have a gold-medal-winning tea called Zhi Lan Xiang, made from one single 300-year-old tree from Wu Dong mountain by Master Lin. When we sampled it, he told us it can be infused fifty times. When we reached the mid-thirties of the sampling, we told him, "We believe you!"'

D'Offay brews a second infusion using the strained Yimu tea leaves. This time the hot water is in contact with the tea for only 20 seconds. 'With this tea, as the leaf has already opened up, the brewing time becomes shorter for the next infusion, whereas with some other teas, if you wanted to make a second infusion, it would need a longer infusion time.' The second infusion oolong is again noticeably fragrant. Tasting it this time, however, the tea has a touch more acidity. D'Offay sips it and nods approvingly – 'I prefer this infusion to the first because there is a little more bite, which to me makes it a bit more refreshing. I like teas where you get sweetness balanced with astringency.' For him the making of tea from leaf tea is part of the pleasure of having a cup of tea. 'I think the making of it prepares you for the drinking of it,' he muses. 'It teaches you patience, to a degree, and it prepares you for the tasting of it, you smell the leaves, you see them. It's not a ceremony, but it is a simple and calming process.'

The following day, working from home at my desk, after a few hours in front of my computer screen, I fancy a cup of tea and head down to the kitchen. Inspired by what I've seen and tasted at Postcard Teas, I don't reach for a tea-bag as has been my habit, day in, day out, for years and years. Instead, I open a packet of Second Flush Darjeeling, bought from them. The instructions are simple: brew with freshly boiled water and allow a teaspoon per cup. After leaving it to steep in the teapot for just a minute, I pour my Darjeeling tea out through a tea strainer into a mug. The tea is bright and clear, a rich golden-brown in colour, with an inviting fragrance. Sipping it I get a delicious grapey flavour with a little trill of tartness at the end. It is invigorating and refreshing – the very qualities for which tea has been valued for centuries.

FAST FOOD

The fast food chain is one of the great success stories of modern enterprise. Outlets selling burgers, pizza or fried chicken dominate high streets around the world. These are international businesses, fine-tuning their global operations to cater for religious requirements – so offering kosher or halal meat – and different national tastes.

The idea of being able to buy something to eat quickly, is, of course, nothing new. Market stalls and street traders have been selling food to be consumed on the spot for thousands of years. Henry Mayhew, a chronicler of nineteenth-century London life, records many foods sold on the street in his great work *London Labour and the London Poor*: 'The solids, according to street estimation, consist of hot-eels, pickled whelks, oysters, sheep-trotters, pea-soup, fried fish, ham-sandwiches, mutton, kidney, and eel pies, and baked potatoes. In each of these provisions the street poor find a mid-day or midnight meal.' As with contemporary fast food, many of these traditional foods were quickly prepared, hand-held and inexpensive: tacos and tamales in Mexico, chaat in South Asia, falafel in the Middle East, porchetta in Italy. In many cultures, this type of indigenous, fast food has a special place in people's hearts.

As a young child living in Singapore in the 1970s I retain vivid memories of visiting 'hawker stalls' at night. Mobile carts containing food and cooking equipment would be pulled to public spaces such as car parks, where they would set up business, selling freshly made dishes ranging from noodles to deep-fried bananas. Standing, watching your order being prepared, was part of the entertainment: the roti canai man deftly swinging out the dough to form a fine, layered flatbread, coals fanned briskly to increase the heat for the satay skewers grilled above them, sugar cane passed through a mangle to crush it and squeeze out the pale green juice. Today in this orderly, affluent city, the hawker stalls have been gathered into centres. Their continued existence, however, and the down-to-earth, affordable food they offer, enjoyed by generations, remains important to Singaporeans. Watching the hawker stalls at work on a recent visit, their efficiency was striking. The diverse elements of the particular speciality being sold are prepared in advance as much as possible: prep done, stock simmered, curry cooked, garnishes chopped, sauces spooned out, ready to be assembled as quickly as possible when an order is placed. Customers are served within minutes, allowing busy office workers to come and eat at hawker centres such as Maxwell Road during their lunch hour.

Still, the dominance of fast food restaurants has seen a backlash. The opening of a branch of McDonald's beside the Spanish Steps in the heart of Rome in 1986 triggered the foundation of the Slow Food movement by charismatic Italian journalist Carlo Petrini. Against 'the universal folly of Fast Life', Petrini posited a defence, beginning 'at the table with Slow Food'. A pressing concern regarding fast food chains is the part they play in the obesity epidemic that threatens the health of so many nations. With their readily available offerings of large servings of cheap food and carbonated drinks, high in salt, sugar and saturated fats, fast food chains are seen as a major contributor.

Britain and other countries, including America, have seen the rise of alternatives to fast food, with the development of a contemporary street food movement. In Britain, its initial roots were in music festivals, with traders travelling round the country, and food markets. Increasingly, cities and towns are seeing street food traders selling in their own right on streets and piazzas, well received by office workers as a lunchtime resource. The classic fast food concept itself has been revised. The simple but profound realization that fast food need not be 'junk food' is at the heart of Leon, co-founded by John Vincent, Henry Dimbleby and Allegra McEvedy in London in 2004. 'We thought,

"Actually there's nothing in the word *fast* or the word *food* that means it has to be bad food,"' Vincent explains. The vision for their 'fast food heaven' was clear from the outset: it was to have the format and size of a classic burger joint and to be found slap bang in the middle of a high street or station, in prime locations. What would be different, however, would be the food, made from fresh, natural, flavourful ingredients, served by people who enjoyed working there. It is not just in Britain that fast food is being reinvented. In the USA, two high-profile chefs – Roy Choi and Daniel Patterson – teamed up to open LocoL in 2016. 'We fundamentally believe that wholesomeness, deliciousness and affordability don't have to be mutually exclusive concepts in fast food,' declares the LocoL website. On a mission to bring healthy fast food to communities living in 'food deserts' – lacking access to fresh food – their first branch opened in the notoriously tough Watts neighbourhood in LA. The menu features cheeseburgers, pizza and milkshakes. Familiar food that people enjoy every day from fast food restaurants, but presented in a healthier form. Fries and sodas are firmly not on offer at LocoL – instead there are sides of rice and 'messy greens' and *aguas frescas*. What was a radical, innovative idea in 2004 – healthy fast food offered with speedy efficiency – is gaining traction.

THE BREATH OF THE WOK

There is a pleasurably vigorous quality to stir-frying, that speedy way of cooking food over a high heat. It involves constantly stirring the ingredients so that they cook evenly, with much satisfying sizzling as ingredients are added to hot oil. This is cooking that engages the cook's senses – sight, hearing, smell – for a short, intense period of time. It is noticeable that recipes for stir-fried dishes give instructions measured in seconds and minutes: 'Add the ginger, stir and add the Chinese broccoli. Sliding the wok scoop or metal spatula to the bottom of the wok, turn and toss in rapid succession for about 1 minute' is a characteristic instruction among the stir-fry recipes in renowned Chinese food writer Yan-kit So's seminal *Classic Chinese Cookbook*.

Stir-frying has its roots deep in Chinese cuisine, although it is a cooking method that has spread widely throughout Asia. 'It developed as a way of cooking because of our history of being a very fuel-poor society,' explains Chinese-American broadcaster and restaurateur Ken Hom. Through his successful TV shows and best-selling books, Hom has done much to spread knowledge of Chinese cooking methods – including stir-frying – to audiences outside China. 'China doesn't have a lot of trees and people needed a way to cook quickly because of the lack of fuel.'

Hom's own memories of stir-frying are rooted in his childhood, when, in addition to watching his mother stir-fry at home, he had experience in professional stir-frying from the age of eleven: 'As a boy, I worked for my uncle in his restaurant, which was, of course, cooking commercially. HUGE woks, and there the cooking time was condensed from minutes to seconds,' he laughs, thinking back to this formative experience in his childhood. Restaurant kitchens – unlike domestic ones – have the advantage of using large, powerful wok cooker ranges, enabling chefs to stir-fry at great speed over high heat.

This brevity in cooking time requires appropriate preparation, hence the careful chopping beforehand of foods such as meat, poultry and vegetables into small pieces that will cook in a short period of time. 'A lot of work is in the preparation,'

Hom acknowledges. The appropriate cooking vessel is essential, too – namely, the wok, with its high, sloping sides, allowing for the food to be moved around inside without it falling out, and, importantly, a small base. 'It's about concentration of heat. It's not a skillet, which has a bigger flat surface that diffuses the heat. When you have a small base, you concentrate the heat, so it gets much hotter.'

Key to successful stir-frying is an intense heat, Hom explains. 'What I advise people to do when they cook at home is to put the wok on a high heat for 5 minutes at least before putting anything in it. When you feel the heat radiating from the wok, *that's* when you add the oil, not before. Remember that stir-frying is done at home in China, not just in restaurants, and this is the method people use – get the wok really, really hot.'

Oil is added and swirled round the hot wok to spread it well, then the hot oil is briefly aromatized with ingredients such as chopped spring onion, root ginger and garlic. There are stages to stir-frying, Hom explains. 'What you have to understand is that you don't stir-fry everything at once. You stir-fry your meat, then you take it out. Then you stir-fry your vegetables, then return the meat and add the sauce.' Throughout the process, the food is kept moving, stirred and tossed to cook it through speedily.

Despite its speed and apparent simplicity, there is a great skill to cooking a stir-fry well. It requires attentiveness, precision, and a genuine understanding of the ingredients and how they will cook. Hom has clear ideas about what constitutes a well-cooked stir-fry. 'A good stir-fried dish should not be oily or greasy,' he says firmly. 'You should be able to taste the freshness of the food. A mistake a lot of people make is that they panic and start adding more oil and that's something you shouldn't do. If what you're cooking is too dry, you add a little water, stock or rice wine. It should also not be watery, though – that happens if you cook the vegetables too long until they go soggy.'

'Because you've cooked in stages, a really good stir-fry will have different layers of flavour and that's what makes it truly wonderful. A good stir-fry will also have overtones of grilled, smoky flavour,' continues Hom, warming to his theme. 'That's why stir-fries are so popular, because of that wonderful smokiness that you get from a hot wok.'

THE 'PERFECT' STEAK

The beef steak has long enjoyed a special cachet. It is an emblematic piece of red meat. Steak symbolizes wealth, virility and luxurious living. Its very simplicity is striking and stylish, satisfying carnivorous cravings in a straightforward way. So popular is it that there are restaurants specializing in serving steak. Resilient in texture, yet juicy within, savoury and succulent, the particular satisfaction a steak offers in terms of mouth feel and flavour makes it a food that people crave. And its appeal is international. Steak plays an important part in countries around the globe, including Argentina, Australia, Brazil, France, South Africa and the USA. It is, however, in Japan – ironically a country of historically low meat consumption – that one finds the world's most expensive beef: Kobe beef from Tajima-gyu Wagyu cattle. The meat from these carefully reared animals is prized for its high level of marbling – the way in which fat is dispersed through the meat – and consequent tenderness and flavour. The fact that there are only around 3,000 Kobe beef cattle means that the rarity of their meat is a factor in the high prices it fetches. So exclusive is it that only a handful of restaurants serve certified Kobe beef steaks. The term 'Wagyu', however, has acquired a wider meaning. Farmers in countries outside Japan, including America and Australia, have begun rearing Wagyu cattle, often cross-breeding Japanese Wagyu breeds with other breeds, but still using the prestigious name.

Steaks, even from 'ordinary' beef cattle, are expensive cuts of meat because they form only a fraction of the yield from a carcass. 'About 20% of a carcass is steak cuts,' fifth-generation butcher Danny Lidgate of Lidgate's, London, told me. The highly prized fillet cut is even choicer, forming only 2–2.5kg of a 300–350kg beef carcass. It comes from a part of the animal that 'doesn't work' (the least used muscle), so is noticeably tender and valued for that reason. How beef carcasses are cut into steaks varies from nation to nation. Brazil's *picanha*, Argentina's *ojo de bife*, Italy's *bistecca alla Fiorentina*, America's flat iron, France's entrecote . . . the list of international steak cuts is a long one, offering much scope for steak geekery. Such is the desire for these prime cuts that butchers struggle to ensure supply. 'Demand for steaks is high and basically that's one of our problems,' commented Lidgate. At Lidgate's, a shop noted for the quality and careful, traceable sourcing of its meat directly from farmers, they buy in whole carcasses. The issue facing them once they've sold the popular prime cuts is how to sell the less popular cuts of meat. The solution, Lidgate explains, is also to source cuts specifically for steaks. Lidgate feels that for his London customers part of steak's allure is the speed at which it cooks: 'Steaks are seen as high quality and quick to cook. I love shin of beef, but a lot of City workers, or people living a fast lifestyle, don't feel that they have the time to do 3 hours' cooking.'

Cooking these costly pieces of meat correctly is an area of

much debate. Order a steak in a restaurant and you are asked, 'How would you like your steak?' – a question linked to cooking times. When it comes to the cooking of steaks, there is a specific, nuanced, sanguinous vocabulary to describe the various levels of done-ness. *Blue* or *bleu* is the term for steak that has been speedily seared on the outside, but is raw within. *Saignant* – meaning 'bloody' in French – is a shade more cooked. *Rare* is still very raw in the middle, while *medium rare* sees the flesh dark pink in the centre. As steak progresses through *medium* to *medium well*, the dark pink flesh in the centre changes hue, becoming lighter in shade as it cooks through. *Well done* sees no trace of red or pinkness in the centre.

Cooking steaks well has been key in creating the success of the Hawksmoor group of steak restaurants in the UK, which opened its first steakhouse in Spitalfields, London, in 2006; their restaurants are where I head to when I want to enjoy a good steak. Serving over 2,000 customers a day, with steak their best-selling item, the group gets through a substantial number. For the group's executive chef, Richard Turner, a no-nonsense, humorous man, there is something special about going out to eat a steak. 'When people go out to celebrate, they'll order a steak and a glass of wine. It has always been a treat. It should be a treat.' With regard to cooking a good steak, the starting point for Turner is the quality of the meat itself. While cross-bred animals have hybrid vigour and will put on weight fast, their meat will lack flavour. 'Flavour takes time,' he says simply. 'Most beef out there is grown with profit in mind, at the expense of flavour.' Hawksmoor, therefore, sources British native breeds that put on weight slowly, looking for cattle that are ideally 36 months old. The age of cattle has become a selling point for restaurants offering steak. Modish London restaurants, among them Kitty Fisher's and Lurra, have acquired a cult following for their intensely flavourful, gamey beef steaks from Galician cattle slaughtered at anything from 8 to 15 years of age.

Time plays a further role in the beef sourced for Hawksmoor, who stipulate that the meat is hung for 5 weeks. Turner feels that 'excessive hanging' is used to cover up the flavour of mediocre beef. 'You can hang beef for 6 months and it would taste funky and gamey. We prefer to age to 5 weeks because this amount of time matures the beef, but doesn't compromise its intrinsic flavour. We know our beef is good, so we want to taste that beef.'

When it comes to how to cook steak, Turner has strong views against both undercooking and overcooking it. 'I think rare is a travesty and so is well done – extremes are just wrong for beef,' he says firmly. People assume that they should automatically order their steak rare, but this is a mistake. Different cuts of steak contain various amounts of fat and so require cooking to a temperature where the fat melts and breaks down, otherwise remaining cold and indigestible. 'Medium rare is the minimum cooking I recommend,' says Turner. When it comes to cooking steak successfully at home, Turner advises that the meat first be brought properly to room temperature (around half an hour out of the fridge), patted dry and seasoned well. The frying pan or griddle should be heated through thoroughly before the meat comes into contact with it. If cooking a fillet, he suggests adding a little fat to the pan, while fattier cuts such as rib-eye don't require it. Once in the pan, the steak should be moved: 'Turn it, turn it,' he exhorts, 'the more you move it, the more Maillard forms.'

When it comes to the vexed question of how to know when your steak is done, advice ranges considerably. Some authorities offer timings given in minutes per side. Other sources advocate feeling the steak, comparing the texture of the meat with different parts of one's hand to judge the resistance. Turner, however, is adamant in refusing to give rigid timings for cooking steaks. The answer to knowing how to cook steaks for the right time is learning through experience. 'Repetition, practice,' is how the chefs at Hawksmoor, cooking over charcoal, are trained, he tells me. 'If you did two days at a grill at Hawksmoor you'd know how to cook a steak; you'd have cooked hundreds,' he grins. 'After years I can now sense when my steak is at a certain level. I'm not entirely sure how,' he muses, 'it's eyesight, but also hearing and smell . . .' It takes time, it seems, to learn how to cook a steak *à point*. Eating a really good steak, butcher Danny Lidgate admits, gives him a 'euphoric feeling, because I understand what's gone into it'. He expands upon what he means. Thirty months of farming 'in all weathers, 7 days a week' – grass-fed beef especially requiring the right cattle genetics, weather and nutritious pasture to allow the animal that life on grass – 'careful, respectful' slaughtering, skilful butchering, proper ageing, further butchering and then the cooking. 'Six stages at which something could go wrong. So, for all of them to come together in a great steak, it's like the planets lining up!'

THE LONG FINISH

As a person who hunts out and enjoys good food, something I've noticed when eating a piece of dry-cured ham or farmhouse cheese is the way in which its flavour lingers on, living on in the palate minutes after the food itself has been swallowed. This aspect of food, whereby the aromas and flavour compounds are still processed and discerned by our brains after eating, has long been prized. The eighteenth-century French gastronome Jean-Anthelme Brillat-Savarin, discoursing on the subject of taste, described the phenomenon with characteristic eloquence: '. . . it is possible to experience a second and even a third sensation, one after the other, each fainter than the last, which we refer to by the words after-taste, bouquet or fragrance; just as, when a keynote is struck, the trained ear can distinguish one or more series of consonances . . .' This long, reverberating finish is a sign of quality, looked for by judges in food and drink competitions, who during the judging process will focus on how the product finishes in the mouth, the changes in flavour and the length of time it lasts.

'I suspect that what is key is, in part, attention,' Professor Charles Spence, whose research explores multisensory perception, tells me. 'The more we concentrate on what we taste, the more salient and, perhaps, longer-lasting, it is to us. One can also prolong flavours by playing with the oral-somatosensory attributes of food.' By this last sentence, Spence means that the bodily sensations within the mouth may affect our perception of its length of flavour, citing the astringency of wine (the drying sensation one can experience when drinking wine) as something he suspects delivers a longer finish. One of the skills artisan food producers possess – with their focus on creating a quality product and their attention throughout the process – is in making foods that deliver this. Even a small portion of their food – charcuterie, sourdough bread, jam – delivers so much flavour, such a depth of taste experience, that one needs very little in order to feel satisfied. Similarly, in the world of alcohol, producers of wines and spirits seek to create this depth and connoisseurs appreciate its presence. The British wine writer Jancis Robinson observes that whereas a mass-produced wine will offer an instant hit of a striking smell, once drunk its impact will quickly fade. In contrast, she writes, 'a mouthful of a great wine tends to last much, much longer'. It is part of the experience of eating and drinking – transient by its very nature, but lingering pleasurably at the best of times.

THE WASABI WINDOW

Traditionally, wasabi is eaten as an accompaniment to Japanese dishes such as sashimi, sushi and cold soba noodles. Wasabi, when I first encountered it in a Japanese restaurant in London, took the form of a small green mound, piled next to my sushi, to which it delivered a pungent kick. This 'wasabi', I know now, was actually made mainly from Japanese horseradish, coloured green to mimic fresh wasabi and sweetened with E420 to counteract the natural bitterness of the horseradish. It is this processed version, sold in paste or powdered form, which has been – and continues to be –predominantly available in the UK.

True wasabi, it turns out, is both rare and expensive. In Japanese culture, the wasabi plant (*Wasabia japonica*) occupies a special place. It has been used for several centuries, mentioned in records dating back to the tenth century, and long valued for its medicinal and antibacterial properties, as well as its culinary ones. It is naturally found growing in and by shady mountain streams and requires 18 months of growth before it can be viably harvested. The prize is the swollen, pale green stem, usually called a rhizome. It is these stems that are grated to produce wasabi paste, eaten as a condiment and appreciated for its particular, refreshing heat. In contrast to the long growing period, fresh wasabi must be consumed within minutes of grating in order to appreciate its aroma and flavour, a fact which limits its commercial possibilities. Undaunted by these facts, however, Jon Old spotted a gap in the UK market and set out to fill it, drawing on his family's experience in aquaculture as watercress growers and co-founding the Wasabi Company, Europe's first commercial wasabi growers, in 2010, to offer own-grown fresh wasabi. At that time, it was barely known in the UK, found only in a few of the most exclusive Japanese restaurants, whose chefs imported it in small amounts from Japan.

The Wasabi Company turned a disused watercress farm in Hampshire into a wasabi farm, planting the first plants in October 2010 in the mineral-rich water of their gravel bed and selling their first wasabi in July 2012. 'Because we wanted to be first to market, we kept the project secret the whole time until then,' says Old with relish. 'We had a code name and enjoyed ourselves enormously.' Fresh wasabi is a luxury even in Japan, with its high price reflecting 'the time it takes to grow, the difficulty in growing it and the fact that all the labour is by hand', explains Old.

My wasabi from the Wasabi Company arrives in an elegantly branded brown cardboard box, a package conveying at once sophistication and Muji-like minimalism. Inside it, I find a carefully wrapped piece of the precious rhizome, pale green in colour with a textured skin. My next step is to grate it in order to sample it. Timings, Jon has explained, are key when it comes to enjoying fresh wasabi at its best, adding a sense of drama to the experience. 'If you slice the wasabi and put it in your mouth, there's no heat and all you get is bitterness. The flavour and aroma of fresh wasabi is a product of the volatile chemical reaction that takes place when you grate it. You want an enzyme that's within the cell wall of the wasabi to mix with a glucosinolate compound inside the cell wall. We say grate, but actually paste it. You want to blast the cells apart; the finer you can get it, the more flavour you get.' Carefully following instructions, I trim the wasabi and, using a circular motion, rub the rhizome over the fine-toothed plastic grater provided, with its texture mimicking the rough sharkskin grater traditionally used for this. Using a dainty brush provided, I then carefully brush the resulting pale green wasabi paste onto a plate to form a little pile. The instructions are to leave it for 5 minutes after grating in order for the flavour to develop. Disobediently, I try some the minute it has been grated; there is a wasabi flavour but very little heat. After the stipulated 5 minutes have passed, I try it again – now there is indeed a fiery heat, that prized 'wasabi kick', but also a fresh grassiness to the wasabi flavour and, to my surprise, a delicate sweetness. It is a subtle, complex flavour, very different from the forceful processed version. As I leave it standing the flavour ebbs away, however, and after an hour it tastes simply mild and grassy.

Despite fresh wasabi's rarefied nature and its novelty in Europe, the Wasabi Company has discovered a receptive audience among restaurants, both Japanese and European, for this expensive flavouring. 'The demand is great. The public are starting to ask for fresh wasabi to go with their high-quality sushi and sashimi. A new production area comes on line this autumn.' Old laughs ruefully, as he considers the logistics. 'We can't grow it fast enough!'

THE ART OF DEEP-FRYING

The French gourmet Brillat-Savarin wrote that 'the whole merit of frying consists in the *surprise*'. This is especially so in the case of deep-frying, where ingredients are plunged into hot oil, undergoing a brief, intense, transformative experience. Deep-frying traditionally confers a new texture on to the food being cooked in this way, meaning that deep-fried foods are at their best when freshly cooked – served while warm, before that desirable crispness created by the process softens in the cooling process.

Though deep-frying is a brief, effective process, it does require patience at key stages. To begin with, there is the initial waiting for the oil or fat to get sufficiently hot. When it comes to testing the temperature of the oil, using a kitchen thermometer is the most efficient way. For those without, one helpful instruction, often seen in recipes, is a time-based one. In her *New Penguin Cookery Book*, for example, Jill Norman writes: 'Test the temperature by frying a cube of bread; if it browns in around 40 seconds the oil is ready. If it takes longer, let the oil heat more. If it browns more quickly, reduce the heat.' Patience is also required if one is deep-frying a substantial quantity of food. In this case, one needs to fry in small batches, allowing the oil to return to full, bubbling heat in the meantime. Overcrowding the pan – while tempting – is a misplaced short-cut as it will bring down the temperature of the frying medium, with sadly soggy results.

With the global success of McDonald's, French fries have become an important part of the fast food experience. Each year, the chain serves nearly 5 million kilos of these. Widely available, bought quickly and cheaply, eaten rapidly, often while on the move, French fries are taken for granted. In fact, the skills required to deep-fry well are amply displayed in the case of French fries. Made from raw sliced potatoes – with no coating involved – one problematic aspect is how to cook the vegetable through sufficiently to be tender without destroying its texture in the meantime. If the French fries are fried so much that they turn dark brown, they develop bitter flavours, as the simple sugars within the potatoes caramelize and burn. One popular approach to cooking French fries is to double-cook them. This involves a first, slow, gentler cooking, for around 6–8 minutes in oil at 284°F, so that the potato pieces are cooked through but barely take on any colour, after which they are drained. The second cooking is a shorter, fierce affair in which the French fries are fried at a higher heat, 374°F, for a few minutes until golden brown, then whisked out, drained and promptly served. The desire to create the 'perfect' fry led British chef Heston Blumenthal to embark on a characteristically obsessive quest. In his much-imitated 'triple-cooked French fries' method the potatoes are first simmered in water until just tender, dried, cooled, then chilled in the fridge. They are then fried at a temperature of 266°F, removed, drained, cooled and chilled, then fried a second time with the oil at 374°F until they go golden-brown. The results of this meticulous, laborious method, Blumenthal promises, will be French fries with 'a crisp crust and a fluffy centre'. Interestingly, the famous McDonald's French fries – mass-produced on an industrial scale with great efficiency – are made using an initial blanching in water for 15 minutes, followed by frying, freezing, then frying again. The time and effort to triple-cook French fries is, it seems, worth it at the upper and lower ends of the food chain.

PASTA PERFECTION

Affordable, easy to prepare, quick to cook and filling enough to form a substantial meal in its own right, pasta is popular across the globe. Defined succinctly by the writer Harold McGee as 'boiled grain paste', this simple food, which comes in numerous forms including noodles and sheets, has an extraordinarily complex history. All accounts agree, however, that two countries – China and Italy – are especially important in its story.

It is the Chinese who are credited with originating pasta many centuries ago. One important piece of evidence for this is an evocative ode written by the scholar Shu Xi (264?–304?) celebrating the making of *bing*, a generic term for dishes made from kneaded wheat flour. In the poem, Shu Xi describes white flour being sifted, kneaded with water or broth until it becomes shiny, filled with finely chopped meat, seasoned with ginger and spices.

Then water is set to boil over the fire.
Waiting for the steam to rise,
We hitch up our clothes, we roll up our sleeves,
And we knead, and we shape, and we smooth and we stretch,
Finally the dough detaches from our fingers,
Under the palm it is perfectly rolled out in all directions.

In Chinese cuisine, two forms of pasta are predominant: the filled pasta shape (such as the wonton dumpling) and the long slender noodle, formed by rolling out and cutting pasta dough or stretching it into thin strands. With characteristic ingenuity, the Chinese did not use solely wheat to create pasta, but also starch from other grains (among them rice) as well as legumes and tubers. Cellophane or glass noodles, for example, are made from mung bean starch. Becoming translucent once immersed in liquid, these fine, wiry noodles retain a resilient texture, very different from more fragile rice noodles. Chinese noodles also range in terms of their width, thickness and in additions, such as egg or flavourings such as seafood or chicken. The hand-stretching of long noodles, such as the wonderfully named dragon's beard noodles – created by deftly swinging strong, supple dough by hand – is an admired skill in China. The results are appreciated for their special silky texture. Food writer Yan-kit So recorded a vivid eyewitness account of a noodle master at work: 'The action from dancing [stretching] the dough to splitting it into noodle strands takes a consummate master about fifteen minutes, but it takes him about two years before he succeeds in harnessing the spontaneous energy to perform it.'

In the West, it is Italy that has a longstanding history of making pasta. Indeed the word itself comes from the Italian for 'dough' or 'paste'. A fourth-century bas-relief in an Etruscan tomb in central Italy depicting utensils similar to those used for making pasta is often cited as evidence of its existence in Italy at that time. The Arab geographer Al-Idrisi, writing in the early twelfth century, described how he saw people in Sicily making flour threads, called *itriyah*. The first written reference to macaroni (thought to be a generic word for pasta at that time rather than the shape we know today) is recorded in Genoa in 1279. In 1356 the Italian writer Giovanni Boccaccio, in his famous work *The Decameron*, wrote exuberantly of a fantastical country called Bengodi, where people living on a mountain of grated Parmesan cheese did nothing 'but make macaroni and ravioli and cook them in capon-broth'.

One of the characteristics of Italian pasta is the use of durum wheat – a hard wheat – to make it. Durum semolina (finely ground, granular durum wheat) is traditionally used for dried pasta in Italy. Its high gluten content ensures the firmness prized in pasta in Italy. The making of pasta from durum semolina is labour-intensive, requiring much kneading to get the dough to the right consistency. During the seventeenth century key inventions – among them kneading machines and the extrusion press (through which the pasta dough is forced, creating a shape in the process) – enabled pasta to be produced mechanically. Over the next centuries the industrialization of pasta-making progressed. In 1933 Braibanti patented the continuous pasta press, which could mix, knead and extrude pasta automatically without stopping. The mechanization of the entire pasta-making process – mixing, kneading, extruding and, importantly, artificial drying – made dried pasta much more affordable and available. In the twentieth century commercially produced dried pasta became a staple food eaten throughout Italy, rather than in the south as had been the tradition. The fundamental part that pasta plays in Italian life is demonstrated by the fact that in 2012 the Italian consumption of pasta was 26kg per capita, more than ten times that of the UK in the same year.

Italian pasta is found in a glorious array of forms. Step into a supermarket in Italy and at least one aisle is devoted to dried pasta in various shapes and sizes: tiny pasta shapes – rings, alphabet letters, grains, squares, stars – for use in soups, curly fusilli corti, long spaghetti, in various thicknesses, farfalle (butterflies), orecchiette (little ears) . . . The richness of choice is in part a reflection of the regionality that characterizes Italian cuisine. It is also a sign of how widely pasta is eaten and how nuanced appreciation of it is within Italy. As any Italian knows, different shapes have different uses, often matched with specific sauces. Fragile-textured, fine cappeli d'angelo (angel's hair) are best in broths, tubular pasta such as penne are an excellent vehicle for meat sauces or béchamel, while in Liguria flat trenette are traditionally coated with olive-oil-rich pesto. With fresh pasta, also, the same discriminating judgements apply. The porous surface of narrow ribbons of fettuccine pairs well with creamy-textured sauces, famously in fettuccine all'Afredo, in which, in Marcella Hazan's recipe, the noodles are simply tossed with cream, butter and grated Parmesan. Pappardelle – the widest of the pasta strips – are served with rich, substantial ragùs made from game or porcini mushrooms. In Italian cuisine the pasta – whether fresh or dried – is considered as important and pleasurable as the most luxurious sauce added to it.

But the industrial manufacturing of dried pasta is often a notably speedy process, with pasta made, extruded and dried in a few hours. Within commercial manufacturing, there is a range of approaches. 'Dried pasta is such a simple, unsophisticated product, and yet there is a huge difference between poor-quality pasta and pasta that has been carefully made and has real flavour,' observes the Italian chef Giorgio Locatelli. A visit to Abruzzo gave me a chance to learn more about the work of a pasta company called Rustichella, one of a number of artisan dried pasta producers for which this verdant, rural region of Italy has long been noted. Produced on a small scale, the pasta Rustichella makes is noted for its flavour and texture, and is used by chefs in upmarket restaurants from North America to Singapore. A family business now run by Gianluigi Peduzzi, it was founded in 1929 by his grandfather Gaetano Sergiacomo. He was a miller in the aptly named, historic town of Penne, and used the flour he ground to make pasta, drying it in the sun, as was the custom then. The drying of pasta in the open air, Peduzzi explains, took place where the weather conditions were right, requiring not only the sunshine, with which Italy is blessed, but also a wind. 'It was a seasonal product,' observes Peduzzi, 'made in the summer months.' Historically, making naturally dried pasta in Italy was associated with specific cities and towns, among them Naples, with its winds from the mountains and the sea. There is a Neapolitan saying: 'Macaroni are made with the sirocco and they are dried with the tramontana.' The sirocco is a hot, humid Mediterranean wind, the tramontana a cooler, drier wind, and both played their part in creating the perfect environment in which to dry pasta, the most delicate stage of pasta production. If the pasta is dried at too high a heat, it risks becoming brittle and cracking. Too slowly, however, and there is the risk that the damp pasta will go mouldy. In Naples, this natural drying process included three stages: the *incartamento* (stiffening), the *rinventimento* (recovery) and the *essiacuzione definitiva* (final drying), a time-consuming stage taking days for long pasta. The temperature at each stage was adjusted accordingly. Each stage required fine judgement over issues such as the thickness of the crust formed on the surface of the pasta and the internal moisture content. At Rustichella, they look to the traditional southern Italian way of making pasta for inspiration. 'What we do today,' Peduzzi tells me, 'is seek to reproduce in factory conditions the way in which my grandfather made his pasta.' The wheat used by Rustichella is both locally grown and imported from Canada: 'Italian for flavour, Canadian for strength.' Taking time throughout the process is key, he feels, to the quality of what they produce.

The Rustichella factory, picturesquely located among groves of olive trees in the Abruzzesi hills, is a small, friendly place. Kitted up with a protective white coat, hair net and shoe covers, I venture inside the production area. As I've been warned before entering, it is indeed very warm and humid, the conditions required to produce pasta successfully. The external weather also has an impact here. Fluctuations in temperature and humidity during March, September and October make them challenging months for production. In July the air is filled with a pleasant nutty smell. On the spick-and-span factory floor, a number of large machines are at work. I watch in fascination as freshly mixed pasta dough is extruded through traditional bronze dies, with the soft, rough metal giving a desirable, sauce-retaining texture to the pasta that is emerging, painstakingly, to form long strands of bucatini. Forcing the pasta dough through bronze, I am told, is slow in comparison to the smooth, slippery Teflon dies used in industrial pasta manufacturing. Strands of spaghetti are the next shape to catch my eye: neat curtains of long, pale golden strands caught on rods and slowly carried through the machine to be briefly 'pre-dried', in order to set their shape. The racks of freshly made pasta, whatever their shape, are then taken to large drying rooms, through which warm air is circulated, drying it for 36 hours. The drying temperatures are kept low, between 104°F and 122°F, alternating periods of heat and rest to allow the humidity to rise to the surface of the pasta. This slow, gentle drying process, taking place at low heat, is evidenced by the pale colour of Rustichella's pasta, contrasting notably with the dark, 'cooked' colour of industrially produced pasta dried quickly at higher heats of 185–203°F. Rustichella works on a far smaller scale than the major Italian pasta manufacturers, producing in a year what it would take Barilla merely 2 days to produce. Peduzzi, a thoughtful, modest, shy man, feels that the results of the time-consuming, careful way in which Rustichella make their pasta are manifest when it is cooked and eaten. 'You taste the difference between artisan and industrial dried pasta; the slow-dried process begins a micro-fermentation that gives flavour, like sourdough bread. I hope you understand there's more than pasta in our brown paper bags,' he says, with palpable pride in his Abruzzesi heritage and the work of his family business.

The time taken by Rustichella to produce their pasta contrasts with the few minutes required to cook it. The speed with which pasta can be cooked is surely a key factor in its international popularity. As the mother of a young son, I was always grateful to be able to quickly cook up a meal of pasta with tomato sauce or simply butter and cheese for him and his friends. The appeal of pasta's rapidity is, of course, exemplified by the large market for instant noodles, invented in Japan in 1958. Pre-boiled, pre-flavoured, dehydrated and flash-fried, to remove any moisture and extend their shelf-life, they are sold, sealed in plastic cups or bowls, to be prepared simply by pouring hot water over them. Such is their popularity around the world that in 2015 the World Instant Noodles Association estimates that nearly 98 billion servings were consumed. It is a sign of the times that Rustichella, looking forwards as well as backwards, now produce Rapida, a pasta that can be cooked in simply 90 seconds, offering nearly instant pasta gratification.

The transformation of pasta from raw to cooked requires immersing it in boiling water, watching it swell and expand as it soaks up the water, softening in the process. Cooking times vary with factors such as size, thickness and shape, and whether the pasta is fresh or dry. Slender, fresh Chinese noodles require only a very brief dip in boiling water to soften them. Dried noodles, in contrast, need longer. Cooking pasta well and for just the right time requires attention. The pasta must be cooked through, yet have resistance when bitten – as the famous Italian phrase 'al dente', 'to the teeth', expressively suggests. In short, it must not be overcooked. This is now a culinary truism. For centuries, though, pasta in Italy was, in fact, cooked until soft, the final triumph of the briefer, firmer style being attributed to the influence of Naples in the years after Italy's unification. The great chronicler of Italian cuisine Pelligrino Artusi, writing in the late nineteenth century, observed that the Neapolitans boiled macaroni 'with plenty of water and not for too long'. The precision with which pasta should be cooked is shown by the fact that De Cecco, a well-known Italian producer of dried pasta, gives two cooking times on its packets: *al dente* and *cotto* (cooked). The best cooked dried pasta I've ever eaten has indeed been in Italy, whether cooked by Italian friends or enjoyed in trattorias and restaurants. When properly done, the firm texture of the pasta, contrasted with a moist, flavourful sauce, is deeply satisfying. For a particularly memorable meal on my Abruzzo trip we visited a restaurant by the Adriatic, to eat Rustichella spaghetti cooked 'alle vongole' as we sat looking out at the beach and the blue waves breaking on it. It is not a complex dish. Fresh clams cooked with garlic, olive oil, white wine and parsley are tossed with spaghetti and served promptly. With its satisfyingly firm texture and discreet, yet pleasant, taste, the spaghetti strands were the perfect foil to the succulent, juicy clams tangled among them, offering their particular salty-sweet flavour of the sea. It is a dish I yearn to eat again. As a friend of mine who has lived in Rome said simply, 'The Italians dare to be brave when they cook pasta.'

THE POWER OF THE PAUSE

We often think of cooking as a series of busy tasks involving heat and motion, yet in fact there are often peaceful pauses, where little is required of the cook, but which, nevertheless, contribute meaningfully – and beneficially – to the final results. Often, I've noticed, culinary pauses are presented in terms of half an hour. Thirty minutes is, it seems, a useful amount of time – simple to remember, easily measured.

There are a variety of practical purposes to these periods. One simple but effective use of this time is to bring something that is fridge-cold to room temperature, uninspiringly referred to as 'de-chilling'. For meat recipes that require browning, ensuring that the meat is no longer fridge-cold before beginning to cook it is a useful step. When frying, adding cold, wet meat straight from the refrigerator to the pan causes the temperature in the pan to fall below the point at which the desired Maillard reaction can happen. Consequently, the meat stews in its own juices first, rather than browning thoroughly and evenly as desired. Cheeses definitely need at least half an hour outside the fridge before consumption, the warmth allowing them to release their aromas, which are such an important part of flavour. The difference between a piece of good cheese eaten fridge-cold or eaten when it has been allowed to reach room temperature is appreciable. Another recommended pause is to 'rest' freshly roasted meat in a warm place for 15–30 minutes after it has completed cooking. This allows the meat to evenly reabsorb the juices expelled during cooking. Food writer Hugh Fearnley-Whittingstall tested the effect of resting and wrote about its benefits convincingly in *The River Cottage Meat Book*: '. . . in the rested joint the equivalent slice was something different altogether: more succulent, juicier and simply nicer to eat.'

An instruction often seen in pastry recipes is to wrap and rest the pastry for 30 minutes in the refrigerator after it has been made, before rolling it out and using it. Often the reason for this step isn't explained, merely given as a slightly enigmatic instruction. There are two main reasons why this step is important. Pastry recipes involve incorporating fat into flour. Chilling the pastry after making it allows the fat to re-solidify, making it able to better hold its shape. As expert baker Dan Lepard notes, pastry 'is easier to roll, shape and fold when it's firm, cool but flexible than when it's warm and soft'. The second reason is to allow the gluten (a strong stretchy protein, aptly named after the Latin word for glue) within the pastry to relax. As Delia Smith explains, the gluten 'develops in time, becoming more elastic and pliable'. Rested pastry, therefore, is easier to roll out finely without tears appearing. Taking time to pause has value in the kitchen.

THE POWER OF STEAM

Steam, created when water reaches a temperature of 212°F and begins to boil, is a formidable force, harnessed during Britain's Industrial Revolution to work engines and drive locomotives. In cooking, the considerable energy taken to transform water into vapour is imparted by steam to the ingredient being cooked, making it a rapid, efficient process.

The lack of direct contact between the hot water and the food makes it an excellent cooking method for fragile ingredients. Vegetables benefit from being steamed rather than boiled, as cooked this way they retain both more texture and more nutrients than when immersed in hot, turbulent water. In Japan steam is used to make *chawanmushi* – a savoury, *dashi*-flavoured egg custard with a noticeably delicate texture – and in China to cook fish, such as prized sea bass, flavoured with ginger and spring onion. That the cooking time for a whole fish weighing around 2.2lb is around a mere 8 minutes shows just how powerful steam can be. In many cuisines, steam is used to cook dumplings, as a visit to a Chinese dim sum restaurant, with its stack of bamboo steamers, each fitted one upon the other, will demonstrate. Steam's characteristic mixture of force and delicacy enables a range of textures to be successfully created, from soft and spongy to slippery and firm. One particularly ingenious creation is the Shanghai-style *xianlongbao*, made by wrapping a mixture of minced pork and jellied stock in fine skins to form pleated dumplings, steaming them for around 8 minutes until the filling has cooked and the stock has melted, then serving juicy 'soup dumplings'. In Moroccan cuisine, steaming is the way in which couscous, a staple ingredient, is cooked, a method that allows the grains to retain their lightness. Traditionally it is cooked in a perforated *couscoussier*, which fits snugly over a pot containing a simmering stew. The cooking of couscous, as recorded by Paula Wolfert in *Moroccan Cuisine*, is a patient, meticulous process, involving washing, then repeated steaming and drying stages. She writes that the resulting couscous will be perfect: 'soft, light, separate, not lumpy, not sticky, not heavy, even though swelled by steam and cold water'.

Britain's steaming tradition is rather less ethereal, used as a way of cooking substantial savoury and sweet puddings. Historically, British puddings, first encased in animal guts, then in pudding cloths (an early seventeenth-century breakthrough), were boiled rather than steamed, but from the eighteenth century onwards boiling *and* steaming became popular cooking methods. The large size of these classic puddings, each one designed to feed a number of people, requires extensive cooking time; Delia Smith's recipe for that classic British dish, steak and kidney pudding, contained in a suet crust, needs 5 hours.

These traditional steamed puddings – Sussex pond pudding, cabinet pudding, jam roly-poly – have declined considerably in popularity. Today it is really only the Christmas pudding – customarily made in advance on Stir-up Sunday (the Sunday before Advent) at the end of November, cooked by steaming for several hours, then cooled and stored – that is still widely consumed. Just a few years ago, steaming the Christmas pudding for hours once more to heat it through would have been part of the Christmas meal preparations; today, however, Christmas puddings can be microwaved in minutes. Ironically, for a nation that was an early pioneer in harnessing the power of steam, our patience with it in the kitchen seems to have become limited.

BURNT OFFERINGS

For thousands of years, cooking food for too long until it begins to burn was taken as a sign of incompetence in the kitchen. As in the story long told to British schoolchildren of King Alfred burning the cakes he was left to watch, it implied lack of attention and carelessness on the part of the cook. The slogan for the online Burnt Food Museum reads: 'Celebrating the art of the culinary disaster'. Nowadays, though, what was once viewed as a mistake has acquired a certain culinary cachet. In modish restaurants around the world it is not unusual for diners to find themselves eating a dish featuring burnt hay, charred vegetables or ash-coated ingredients.

As with many contemporary food trends, Spanish chef Ferran Adrià of El Bulli seems to have been a major innovator, offering a dish called 'vegetables on the grill' during the late 1990s featuring charcoal oil. René Redzepi of Noma in Copenhagen has also played with burning food in characteristically imaginative fashion. Recipes in his monumental *Noma* cookbook reveal him charring meringue pieces with a blowtorch until darkened, and coating cooked leeks with a burnt leek purée. 'Sprouts, like all cabbages, take well to a little charring,' observes Yotam Ottolenghi, another chef noted for setting trends, in his weekly recipe column for the *Guardian*. His recipe for burnt Brussels sprouts with cream cheese and ginger dressing stipulates cooking the halved sprouts in a very hot, covered pan – with no oil – 'until blackened'. The words 'burnt' and 'ashes' appearing on an elegant menu have a frisson, evoking the elemental power of fire. 'The bitterness imparted to ingredients through burning appeals to grown-up, sophisticated palates. Today's use of ash reflects a playful creativity in gastronomy, a pushing of boundaries by chefs for whom innovation is the new orthodoxy. A reason for its popularity is partly that the visual power of burnt food is huge. Blackened ingredients, whether lettuces, onions or clementines, bring an element of drama to the plate. In an age of social media – where a striking Instagram #foodporn image can do so much to promote a restaurant – one can understand why burning has cult appeal.

But ash has, in fact, been used in certain cuisines for generations, notably in Central America. The ancient process of nixtamalization (from the Aztec-language word *nextlli*, for ashes) dates back thousands of years in Mesoamerican history. Ash is used in the soaking and cooking of corn (indigenous to that part of the world) because its alkaline properties break down the thick, tough outer pericarp that encases each kernel by dissolving the hemicellulose (that acts like a glue) in the plant cells' walls. Not only does nixtamalization make the corn far easier to grind, it also gives it flavour. Usefully, too, the process releases vitamins such as niacin (vitamin B3) stored within the endosperm and germ, making them digestible by humans. In Europe, ash was historically used to coat cheeses, helping to preserve them by creating a layer on the rind. This usage continues in French cheeses such as Selles-sur-Cher or Valençay, the black coating contrasting with the bright white paste of the goat's milk cheese inside.

MAKING MAYONNAISE

Mayonnaise – that voluptuous, glossy, smooth-textured, cold, emulsified sauce made from egg yolks and oil which has a special place in French cuisine – retains its aura of an indulgent treat, especially when home-made. Traditionally, mayonnaise is made by hand using a wooden spoon or whisk, a patient process in which the oil is added to the yolks a little bit at a time. What gives mayonnaise its special texture is the fact that it is an emulsion, a mixture of tiny fat particles suspended in water. Normally, oil and water separate themselves, so to create the emulsion requires both emulsifiers – found in the egg yolks – and a process of mixing them together: the vigorous whisking. The transformative aspect of making mayonnaise – watching egg yolks and oil change before your eyes over a period of minutes into a thick, viscous, opaque substance – has a special satisfaction to it.

The difficulty in achieving a successful emulsion means that there is a certain mystique to the making of mayonnaise by hand, with the ability to do so conferring a culinary cachet on cooks. Julia Child, writing on mayonnaise in *Mastering the Art of French Cooking*, tells her readers firmly, 'you should be able to make it by hand as part of your general mastery of the egg yolk'. Elizabeth David, in *French Provincial Cooking*, writes with characteristic eloquence of the pleasures of making mayonnaise as opposed to using a liquidizer: 'I do not care, unless I am in a great hurry, to let it deprive me of the pleasure and satisfaction to be obtained from sitting down quietly with bowl and spoon, eggs and oil, to the peaceful kitchen task of concocting the beautiful shining golden ointment, which is mayonnaise.' Classic recipes for hand-beaten mayonnaise, including those by Julia Child and Elizabeth David, follow the same pattern – first the egg yolks are beaten well, then the oil is added very slowly initially, 'drop by drop'. Once the emulsification process begins, however, the oil can be added more quickly. The risk when making mayonnaise is that it will turn and not thicken or, having thickened, that the oil will release itself from the mayonnaise, causing it to curdle – a most dispiriting experience. In that event, it is necessary to begin the process again with an egg yolk in a clean bowl, gradually incorporating the failed mayonnaise.

British chef Rosie Sykes is someone whose knowledge of cooking I have long admired. I am pleased to learn that she enjoys using home-made mayonnaise in her own cooking, valuing its versatility. It is often a base sauce, to which she adds, for example, mashed egg yolks, finely chopped egg whites, tarragon, parsley, capers and cornichons to make a tangy sauce gribiche, to be served with asparagus or used to dress potatoes, or which she flavours with anchovies and garlic for a Caesar salad dressing. When making mayonnaise by hand, she estimates the time needed to be around 10 minutes, 'But I whisk for a living!' she points out, feeling it would take a home cook rather longer. Sykes's own recipe for mayonnaise breaks from classic yolk-only versions in using a whole egg plus a yolk, with her reason for including an egg white being that 'it makes a much lighter mayonnaise, that's more stable'. She also recommends using a food processor with either the plastic or metal blade, preferring the resulting texture to that made by hand-whisking. Although using a food processor takes away the work of hand-whisking, her method is still a gradual one. First, she stipulates, the egg and yolk should be very fresh and at room temperature. The next important thing is a generous tablespoon of Dijon mustard – added to the egg and yolk and then whizzed together in the food processor for a few minutes until well mixed. For that quantity of egg, she suggests 8.45 fl oz of sunflower oil, with 1.7 fl oz of light olive oil, feeling that using all olive oil is 'incredibly rich. You have to have the right things to eat it with.' With the food processor running, the next step is to put the oil in a jug and pour it in in a very slow stream. 'To start with, I would go *really slowly*,' Sykes emphasizes, 'because that's how you're going to get your emulsion. Add a little, say 2–3 tablespoons of oil, then leave it for a few seconds. Do this about five or six times, then you get to the stage where you can add the oil in a steady stream. You'll see that the mayonnaise begins to emulsify, it puffs up and changes colour, goes a lot paler. Fergus Henderson has this wonderful video, *Listening to Mayonnaise*, showing how you can tell how your mayonnaise is working in the food processor by the changes in noise. It changes from being really watery to a slapping noise.' Once all the oil has been incorporated, Sykes tastes it and seasons with salt and freshly ground pepper. The texture at this stage is distinctly thick, so she adds an acidifier, such as lemon juice or vinegar, in order to thin it and make it more spoonable. Sykes, it is clear, has a very particular idea of what she looks for in mayonnaise, in terms of both consistency and flavour. 'I think most people are pretty happy with the mayonnaise I make,' she says reflectively.

SPEED-EATING

Seventy-three grapes eaten in 1 minute. Twelve hamburgers eaten in 3 minutes. Six hundred and sixty-three grams of jelly eaten in 1 minute. Simply reading these entries in the *Guinness Book of Records* is enough to give me indigestion. Eating contests, focused on who can eat the most, quickest or within a given time frame, trace their origins to country fairs during the nineteenth century. They became a popular form of entertainment during the twentieth century. Doughnuts, pies, pasta were among the foods popularly scoffed. These days the range of foods competitively consumed includes challenging ingredients such as formidably fiery Bhut Jolokia chillies or live cockroaches. One of the most iconic food-eating competitions is Nathan's Hot Dog Eating Contest, held on Coney Island on the Fourth of July. Said to trace its origins back to 1916 as a contest among immigrants to demonstrate their patriotism, the competition has grown hugely in popularity since the 1970s. Contestants are given 10 minutes in which to eat as many hot dogs plus the buns as they can, with the current champion being nine-time winner Joey 'Jaws' Chestnut, who consumed a staggering seventy. Consuming so much food so fast puts one's body under considerable pressure. Our usual satiety reflex that tells us when we are full has to be overcome. Vomiting is a natural response to excess consumption, but doing so during or immediately after competitions results in disqualification. Choking is a very real danger: in November 2016 a Japanese man choked to death on a rice ball during a speed-eating competition. The long-term consequences of competitive eating down the years – what repeatedly stretching one's stomach to such an extent might do to it – are as yet unknown.

Despite its risks, competitive eating is on the rise. What were once sideshow events, held for fun by amateurs, have now become big business, with male and female 'professional' eaters competing for considerable prize money in events run by Major League Eating, a self-styled 'world body that oversees all professional eating contests'. Contestants train in advance, stretching their stomachs in preparation. Interviewed about his preparation for Nathan's, Chestnut describes how his training includes drinking a gallon of water first thing in the morning, downing it in as few gulps as possible. The first few minutes after a contest are the worst, he says, with the fear that every burp might turn into regurgitation and result in disqualification. 'I'm pretty much drunk on hot dogs. I feel like there's something wrong with me,' he says simply. Time in the world of speed-eating becomes an antagonist – something to do battle with, whatever the physical consequences.

FLAVOURFUL FRYING

Flavourful frying – an all-important initial stage that gives taste and aroma to the dish being made – is found in cooking across several cuisines. Many Chinese dishes start with hot oil being aromatized by chopped root ginger or garlic thrown into it and briefly fried until fragrant before other ingredients are added to the oil. In French cuisine the sautéing of a *mirepoix* (a diced mixture of vegetables, classically onion, carrot and celery) is the initial step in dishes such as braises, stews and sauces. Beginning a dish with a *refogado* – made by frying ingredients such as onion, garlic, leek and green pepper – is an essential step for many Brazilian dishes. In Italian cuisine *battuto* is the name for a finely chopped mixture of flavoursome ingredients, usually containing onion, sometimes with the addition of carrots, celery and fresh herbs such as parsley. When gently fried until the onion becomes translucent, this becomes the *soffrito*, from the Italian verb *soffrigere*, meaning 'to sauté'. Marcella Hazan, the renowned, authoritative Italian food writer, is clear on the importance of this preliminary frying stage. 'An imperfectly executed *soffrito* will impair the flavour of a dish no matter how carefully all the succeeding steps are carried out,' she writes with characteristic firmness.

In South East Asian cuisine the starting point for many dishes is an intensely flavourful paste made by pounding together ingredients such as shallots, garlic, galangal, lemon grass and spices. This paste – known as *rempah* in Malay – is then fried until cooked through, a process known as *tumis*. Taking enough time to fry a *rempah* properly is important, as, if it has not been well cooked, the resulting dish will have a harsh, raw taste. Intrigued to know more about how to fry a *rempah* correctly, while on a trip to Kuala Lumpur, Malaysia, I visited Jeanne Pereira, a noted Portuguese Eurasian cookery teacher, at her home. Vivacious and charming, and happy to share her culinary knowledge, Pereira kindly demonstrated how she cooks a curry, beginning with the all-important spice paste frying stage. The first thing I notice is the large amount of oil in the heavy-based *kuali* (wok), enough to form a noticeable layer in the base. Having ample oil is important, Pereira emphasizes, in order to prevent the paste from burning as it cooks and stop the sauce from having a gritty texture. After all, she points out, any excess oil can be skimmed off before serving the dish. I watch as first a few whole spices – cardamom and cinnamon – are fried briefly, then a paste of minced onion and garlic is added to the hot oil with a satisfying sizzle and fried for 3 minutes, stirred constantly as it cooks. 'You want to brown the onion a little but not to fry it until it's crisp,' explains Pereira. She throws in a handful of fresh curry leaves. As she fries the mixture, Jeanne uses her spatula to spread the mixture around the sides of the *kuali*, moving it around away from the direct heat, looking to reach the point where it is 'not too dry nor too wet'. Three minutes later she adds in a curry powder paste, mixing it thoroughly into the fried onion and immediately sprinkling in a couple of tablespoons of water and scraping the pan vigorously with her spatula to prevent the spice paste from sticking and burning. 'Make sure you stir the base,' she warns. As she fries the paste, she carries on sprinkling a little water, always just 2–3 tablespoons at a time, cooking the paste until it dries out, then adding a bit more water. As it cooks the paste darkens in colour and takes on a thick, porridge-like texture. Around 5 minutes later, the oil has risen to the surface of the paste, a sign that Jeanne is looking for. Jeanne continues the careful process of moistening the paste and cooking it, always scraping the base thoroughly and effectively. Fifteen minutes after Pereira began cooking, she observes with satisfaction, 'This is done – you see the colour and get the aroma.' She uses one of her hands to wave over to me an appetizing waft of a rich, deep, spicy smell. She proceeds to add chicken pieces and chopped potatoes, gradually stirring in water to transform the fried paste into a thick gravy in which the ingredients simmer until they are cooked through. 'The important thing is the proper frying of the *rempah*,' Pereira emphasizes. 'Curry is something you just cannot rush.' She smiles in recollection. 'When my mother first showed me how to make this I was so bored! All this standing and stirring! Now she's laughing at me from above!'

SEARCHING FOR SOCCA

It was with real excitement that I saw the stall at Nice market. Here, at last, was my chance to sample socca, the chickpea pancake, a famous, historic French Riviera snack which I'd read about but had never yet eaten. On booking our holiday in Provence, a good friend with impeccable taste in food had told me to be sure to try it. We were near the end of our Provence holiday, however, and so far socca had proved elusive. We were staying inland – not the area for socca, I'd discovered. A trip to picturesque Antibes market disappointed when it turned out the usual socca stall was not there that day. Here in Nice, however, was a stall with a sign proclaiming 'Socca'. As I drew nearer, I could see that rather than being cooked fresh, what looked like a large round pancake was being reheated. I bought a portion, tore off a piece, ate it and was disappointed – soft, flabby-textured and lacking in flavour; in a word – dull. My husband and son agreed and were ready to move on from the socca quest. A certain stubbornness in me came to the fore. 'I read of a restaurant, Chez Pipo, where they cook socca in a wood-fired oven. It might be different . . .' My husband looked at me with a certain resigned expression and said, 'If it's not too far away; my camera's heavy.' Half an hour's walk later, tracing our way through bright, hot sunny streets, we found Chez Pipo, tucked away in a side-street. It was a down-to-earth, informal place, busy with lunchtime diners. The menu was short and the choice limited. Twenty-five minutes' wait for the socca, the waitress informed me. I ignored a reproachful glance from my husband and ordered it brightly. At last, a plate was put down in front of me. On it were strips of freshly cooked socca, still warm, golden-brown splotched with dark brown patches, far more visually appealing than the pallid, limp market version. We seasoned it with ground pepper, picked it up and ate it eagerly; the atmosphere round the table brightened noticeably. The pancake was crisp on the outside, soft inside, with a distinctive chickpea flavour – now I could understand socca's cult status. As my friend had correctly told me, it is simple but good. So good, indeed, that we ordered a second portion.

On the way out, my husband paused to peer round a screen, then beckoned me over. 'Come and look. You'll love this.' There was a huge oven, with a wood fire visible at the back, and busy at work was the socca-maker, a young man, who flashed a happy smile at us before turning back to tend the single large pan containing socca in the oven. His name was Bruno and, he told us, he'd been making socca for 7 years. Responding to our interest, he told us more. The impressive 2 feet, 8 inches circular flat pans in which he made the socca were lined with copper on the base in order to conduct the heat efficiently, with the pans needing re-covering every 2 months. The chickpea batter is rested for a period, then spread thinly in the large pans. As he spoke to us, Bruno kept a vigilant eye on the socca as it cooked in the oven, drawing it out and deftly scratching the surface, adding oil if he felt it was too dry. The oven holds only one socca pan, each socca takes 15 minutes to cook and gives only 9 large portions. If we had come in the busy summer months, we would have had to wait even longer for a portion; 2 hours' wait for a taste of socca is usual at Chez Pipo, he laughingly informed us. We left satisfied and happy – pleased to have experienced proper socca and witnessed Bruno's pride and pleasure in his job. It was good to know that, even in our restless age, people are prepared to wait for something to be freshly cooked, and that socca remains special in Nice.

THE MAGICAL MAILLARD REACTION

Many recipes for meat dishes involve frying the meat at a preliminary stage. In the past, often the reason given for doing this was in order 'to seal in the juices', which is, in fact, inaccurate and misleading. What this browning process does do is create both colour, which we find pleasing visually, and, more important, flavour. There are certain cooking processes that it is well worth taking time over, and the browning of meat is one of them. In her much-loved cookbook *Sunday Suppers at Lucques*, Californian chef Suzanne Goin gives a recipe for *boeuf à la niçoise*. First the boneless beef ribs are marinated overnight, then, once brought to room temperature, the instruction is given to sear the meat in batches in a heavy pan in very hot oil until well browned. Tellingly she adds, 'This step is very important and should not be rushed; it will probably take 15 to 20 minutes.' Watch a good chef at work frying meat for a stew or a braise and you will indeed notice the time they take.

As Goin and other chefs know, browning meat in this way is an important step in enhancing its flavour. This chemical reaction which causes foods to acquire flavours and aromas we find deeply appetizing is known as the Maillard reaction, in honour of the French chemist Louis-Camille Maillard, who first discovered the chemical process and described it in a paper in 1912. The first simple reaction between amino acids (the building blocks of protein) and sugars triggers a series of further reactions, creating hundreds of flavour compounds in the process.

There are numerous examples of the Maillard reaction in cookery across different ingredients, not only meat. An everyday example of the Maillard reaction at work is the making of toast, with further examples being the famously appealing smells created when baking bread or roasting coffee. The temperature at which the Maillard reaction speeds up considerably, impacting meaningfully the food being cooked, is around 350°F. This is higher than the boiling point of water at 212°F, which is why boiled, steamed or poached foods do not brown, while foods which are fried, grilled, seared or roasted at higher temperatures do. So complex is the Maillard reaction, however, that it is still not fully understood, although scientist John E. Hodge did important work in 1953 in establishing a model for the reaction consisting of three stages.

From a cook's point of view, it is useful to realize quite how important browning can be in creating flavour. The way in which the humble onion is enhanced through the Maillard reaction demonstrates its power. Many recipes for braised dishes such as curries begin with the frying of onions, often until they have browned. My favourite korma recipe is by my friend the talented Indian food writer Roopa Gulati, and key to its depth of flavour is the making of a deep-fried onion paste. First, the onions are finely sliced, salted, set aside for 30 minutes in order to remove excess water, patted dry, then deep-fried in oil until they go a rich, dark brown. The fried onions are then blended with a little water in order to form a paste and this is added to the chicken, creating a gloriously savoury base for this aromatic dish, fragrant with spices including black and green cardamoms, cinnamon and coriander. A relative amount of patience may be required, but each time I cook this dish I appreciate the considerable culinary rewards of harnessing the magic of Maillard.

GETTING RICE RIGHT

My father taught me how to cook rice when I was 10 years old. A truly useful piece of knowledge to pass down the generations, I feel, and something which I, in turn, have taught my son. As an Englishman born in 1929, my father must have learned how to cook rice this way while living abroad working as a librarian in Malaysia. The first step he showed me was the patient rinsing away of excess starch from the rice. This was done by hand in the sink. I watched the water in the saucepan turn cloudy as I stirred the rice and released its starch, feeling the slender grains with my fingers, a pleasant sensation. The drained rice was placed in a small, heavy saucepan, then covered with water to a knuckle's depth above the rice in the pan, measured with one's finger resting lightly on it, and a pinch of salt added. The water was brought to the boil and, as soon as it bubbled, the pan was covered tightly, the heat turned down as low as possible and the rice left to cook undisturbed for around 12–15 minutes. By the end of the cooking time, a transformation had taken place: the water had disappeared completely, leaving white, fluffy yet textured, dry, cooked rice. As a child, the transformation – with its perfect results – had almost a magical quality to it. He had taught me the classic absorption method of cooking rice – making it so simple and clear and quick that a child could follow it.

It was only when I ate rice at other families' homes that I realized there were other ways to soften the small, hard grains to the point where they could be eaten. One popular way of cooking rice in the England of my youth was simply cooking it in a large pan of boiling water until soft, draining it and serving it. To my critical eye, this overly soft, soggy rice lacked the subtle charm of rice cooked by the absorption method. To this day, this boil-and-drain method is often given as a cooking method on packets of long-grain rice. What is striking is that, while as a method it is a few minutes faster than the absorption method, the results, to my mind, are still distinctly inferior. I am not alone in having decided opinions on the correct way of cooking rice; caring for rice is something found in countries around the world. In rice-loving cultures, the final texture of cooked rice is seen as an important way of gauging how well it has been cooked. The cooking of rice, it turns out, is a complex subject. To begin with, there are many different types of rice, broadly divided into long-grain, medium-grain and short-grain. These types require different cooking times and suit diverse cooking methods. The reason for this is to do with how rice is constituted. Rice is a seed containing starch; when cooked in water the starch granules absorb the water, swelling and softening, a process known as gelation. Different types of rice contain varying amounts of two types of starch – amylose (a simple-structured starch molecule) and amylopectin (a larger, more complex molecule). The starches affect the structure of the cooked rice. The high amount of amylose contained in long-grain rice varieties means that the grains stay separate after cooking and suit dishes such as an Iranian *polow*, where this lightness of texture is sought. Short- to medium-grained types of rice, such as Italian risotto rice and Spanish paella rice, have low levels of amylose and a higher amount of amylopectin, making the rice softer and stickier when cooked and, indeed, ideal for risotto and paella. Sushi rice has tiny amounts of amylose and high amounts of amylopectin, giving the cooked rice its distinctive sticky texture and causing the grains to hold together when formed into sushi.

One classic way of speeding up the cooking time required for long-grain rice varieties is to pre-soak them, so that they have already absorbed some water when you start cooking. In Iranian cuisine, where rice is highly esteemed, with rice-based dishes such as *polows* an important part of a meal, it is taken for granted that the preparation of rice involves soaking it. In her delightful cookbook *The Legendary Cuisine of Persia*, Margaret Shaida describes the preparation process for cooking rice in the Iranian style. First the long-grain rice is washed to rinse off excess starch. Next, it's soaked in salted water. In Iran, 'where the rice is hard and the climate dry, the rice is soaked for many hours, usually overnight'. For Basmati, however, she writes that a mere 3 hours' soaking is required. The rice is then par-boiled and steamed. One of the specific characteristics of an Iranian *polow* is the crusty layer of rice at the base of the dish, created by adding rice into hot fat, and its presence considered a sign that the dish has been properly cooked. Called a *tahdeeg*, it is a delicacy: 'It is rich, crisp and tasty and has a habit of disappearing in the kitchen before it ever reaches the table,' notes Shaida. In Spain, where paella, the national rice dish, is made with short- to medium-grain rice, the crunchy crust of rice formed at the base of the dish, called *socarrat*, is similarly prized. Again, its presence is a sign of correct cooking.

Rice plays a central part in Indian cuisine – a staple crop, widely grown in the country and forming a basis for most meals. Basmati rice, grown in the foothills of the Himalayas, is especially prized for its fragrant aroma. A festive dish that traditionally showcases Basmati to excellent effect is the biryani. Historically associated with Mughal cuisine, again it requires time and care to make it well. Par-boiling of the rice in water is an important stage, before layering it with richly spiced ingredients such as meat, poultry, fish or vegetables. 'The reason why we only half-cook the rice first is that we want the rice to absorb the flavour of the juices and the aroma from the spices as it continues cooking in steam, swelling and softening,' Indian food writer Roopa Gulati tells me. The rice is soaked for 20 minutes before cooking to soften it, remove excess starch and give a whiter colour. The soaked, drained rice is then cooked briefly – for 2 minutes – in boiling water and drained. 'You have to resist the temptation to cook the rice too much,' cautions Gulati. 'If anything, undercook it.' Working quickly in order to capture the heat, the part-cooked rice is then layered with a spiced curry in a deep dish and covered tightly. Traditionally the tightly fitting lid is sealed with chapatti dough to ensure that no steam escapes. The dish is then baked for around 45 minutes at a low temperature, during which time the rice should complete its cooking, becoming tender. The cooked biryani is brought to the table and the lid taken off. This is the moment of truth for the biryani cook. 'What I'm looking for is separate grains – it's one of the ways people judge a biryani – and a wonderful mellow aroma which comes from the melding of all the spices,' says Gulati with relish. 'When you serve it there should be separate layers of rice and spiced meat, and the meat should still hold its shape.'

Short- to medium-grain rice presents different challenges for the cook, famously in Italy's classic risotto. This is traditionally made from specific Italian rice varieties – Arborio, Vialone

Nano or Carnaroli – that are cooked in stock. A distinctive feature of the dish is that rather than the rice being simply boiled in liquid until tender, the stock is added gradually over a number of stages. While it cooks, the rice is stirred frequently to ensure even cooking, so it requires attention throughout. Like many Italian dishes, there is a skill to executing it well. The rice should be cooked through, yet retain bite, while the final texture should be creamy rather than gluey.

For advice on risotto, I turn to someone whose cookbooks I've long enjoyed using, the respected Italian food writer Anna Del Conte, author of the authoritative *Gastronomy of Italy*. 'It takes around 14 to 18 minutes to cook a risotto,' Del Conte tells me. 'You have to be there and be careful. The better the variety the longer it takes to cook a risotto. Arborio is NOT the best. Carnaroli is better.' Del Conte talks me through the process of making a good risotto. First melt butter in a pan and fry onion in it until softened, then add the rice, stirring it to coat it with the butter. The rice should cook in the butter for 2–3 minutes, until the outside of the grains becomes translucent. If using wine, add it at this point and allow it to cook off. Next add some simmering stock. 'The quality of the stock is very important,' exhorts Del Conte. 'Add slightly more with the first go, mixing it in well.' Once this stock has been absorbed, another 150–200ml of simmering stock is added to the rice. 'It's difficult to explain the temperature that one should cook the risotto,' muses Del Conte. 'It shouldn't be bubbling too hard, nor should it be bubbling too low. It's only experience that can tell you.' Pondering what might make a bad risotto, she feels that it would probably be the result of 'cutting corners', such as not frying the onion enough at the beginning or not sautéing the rice adequately. Another fault might be 'adding too much stock all at once, rather than adding it slowly'. A poorly made risotto, furthermore, is one in which the rice is overcooked. 'One minute and it's done. It's like pasta, it should not be overcooked,' she says firmly. 'It should have a little bite inside. The bite is important.'

UNDER PRESSURE

Due to its alarming hisses and whistles and clouds of escaping steam, a stove-top pressure cooker can be a daunting presence in the kitchen. Despite this, its ability to save time when cooking means that pressure cookers are widely used in countries around the world.

At its heart is a simple but ingenious idea: to trap the power of steam, so increasing the pressure inside the appliance. Its ancestor was the steam digester or bone digester, invented by French physicist Denis Papin in 1679, complete with a safety steam-release valve to prevent explosions, and designed to cook bones until they softened. Modern pressure cookers follow the same principles; the addition of a liquid element, usually water, to the raw ingredients is essential to using them. The pressure cooker is then heated until the liquid turns into steam, which, with nowhere to go, increases the pressure, raising the boiling point to temperatures as high as 237°F or 250°F. The high temperatures not only allow for much faster cooking but also for the all-important Maillard reaction to take place, creating flavourful food. The acclaimed British chef Heston Blumenthal is a declared fan of pressure cookers, advocating their use for making stocks on the grounds that the process seals in the flavour molecules, the high heat creates the desired meaty flavours and the lack of turbulence during cooking gives a clearer stock.

When it comes to saving time, the pressure cooker is impressive; a general rule of thumb is that it cuts down on the time taken for conventional cooking by 70%. It is a staple piece of kit in many Asian households, enabling pulse-based dishes and curries that would normally require hours to be made in minutes instead. In Britain, they are less common, but Catherine Phipps, author of *The Pressure Cooker Cookbook*, is championing their benefits when it comes to cooking dishes from risotto to Chinese spare ribs. She had her own 'Damascene conversion' when she watched her Brazilian sister-in-law cook a *feijoada* (a black bean stew) using one in just 30 minutes, a process that would normally take hours, without even having to pre-soak the beans. In Brazil, she was told, they are an everyday kitchen appliance. Intrigued, Phipps invested in a pressure cooker and, as a busy working mother, realized its scope for saving time when cooking. The poor image that pressure cookers have in Britain is, she feels, due to the fact that they became widely used during rationing. 'A lot of the recipes from that time are for boiled hearts and kidney stews, lots of grey casseroles and sludgy stuff!' Phipps points out that using a pressure cooker saves not only time, but also energy and therefore money. Her affection for this unsung piece of kitchen equipment is manifest. 'I strongly believe in cooking, but like a lot of people I don't have a lot of time in the evening and feel tired at the end of a day.' Using a pressure cooker allows her to make a comforting risotto from start to finish in just 10 minutes. 'For me it's a marriage of convenience and cooking from scratch that allows me to make proper meals.'

THE BROWNIE MOMENT

Baking is an area of cooking that famously requires precision. To ensure success, ingredients should be weighed, recipes followed thoroughly, ovens preheated to the required temperature. If not, disappointment awaits. When it comes to assessing the 'done-ness' of a cake – the point at which it is ready to be removed from the oven – the customary test is to insert a skewer into the middle to check that it comes out clean. Any trace of uncooked cake mix on the skewer means the cake requires longer baking. When baking brownies, however, assessing the moment of brownie perfection is rather trickier. Brownies have a very specific soft texture. Pull them out too early and an overly liquid, uncooked centre is a danger. Cook them for too long and the results are disappointingly dry. What I want is something gloriously in-between. Finding this moment, however, can be elusive.

A visit to Outsider Tart in Chiswick in west London reveals the diverse possibilities of brownie-baking. Co-founded and run with cheerful obsessiveness by baker David Muniz and his partner, David Lesniak, the shop-cum-café celebrates the richness of their native baking heritage. The daily display of freshly baked goods – cakes, cookies, whoopee pies – on the counter is an impressive sight. Brownies here come in numerous incarnations: caramel-glazed Snickers brownies, Hepburns, Mile High, blondies. Yet, across the range of flavours and additions, the texture is the same: a moist, yielding fudginess, which I find appealing.

Engaging and softly spoken, Muniz explains his approach to making brownies, first emphasizing that the choice of texture is very much a personal one. How you make the mixture will affect the final texture, he tells me, 'because you get different results based on the temperatures of some of the ingredients you're putting together, the order you put them in and how much you mix. If you over-mix, you get a cake-y brownie. I put the flour in at the very end, gently fold it in and then bake it.'

Getting the timing right is vital. Once the brownies are in the oven, Muniz tells me, he always subtracts at least 10 minutes from what it says on the recipe and starts checking them then. Vigilance in the final baking stages is key: 'I compulsively check during the last few minutes,' he says with a smile. Rather than piercing the mixture to see how cooked it is, Muniz assesses by sight. 'You want to see the mixture slightly pulling away from the side so you know the edges are cooking fine.' Muniz also looks at the middle of the brownie mixture, which might be glossy or matt, depending on how it was mixed and made. 'If it's wet you can always see a darker staining in the centre when it's still wet – you don't have to touch it or stab it – and you keep cooking if you have that darker colour in the centre.' There is a precision, gained from years of experience, to Muniz's description of making brownies. In the final moments, the mixture should be all one even colour. 'When you shake the pan you want it not to move anywhere, but you want to get it when it *just* stops moving,' he says tellingly. The reason for this specificity – the awareness of visual and textural signals – is to achieve the desired thick, soft but not runny texture. He pauses reflectively. 'It depends on what you like. If I want a cake I'll eat a cake, if I want a brownie I want fudgy and rich and moist.'

SPEEDY COOKBOOKS

For much of human history, recipes were passed on orally through generations. Often this process took place within families, with daughters shown how to cook dishes by their mothers or grandmothers. Learning to cook from experience in this way is a sensory method, based on sight, touch, taste, smell and sound, to understand salient points such as the texture of dough or when a sauce is ready.

When written recipes did appear, it is striking how brief and terse they were. One of the earliest known instructive cookery manuscripts in the English language is *The Forme of Cury*, thought to have been written at the end of the fourteenth century by the master-cooks of Richard II (1377–99). 'Cury' was the Middle English word for 'cookery' and the book contains 196 recipes, ranging from everyday dishes such as common pottages (thick soups) to grander fare. The style is pithy and to the point. The entire recipe for pork in sage sauce consists of three sentences. 'Take pigs scalded and quarter them and seethe [boil] them in water and salt, take them and let them cool. Take parsley, sage, and grind it with bread and yolks of eggs hard boiled. Temper it up with vinegar somewhat thick, and lay the pigs in a vessel, cover it with the sauce and serve it forth.' These are recipes as an aide-memoire, written for people who already knew how to cook, simply outlining how a dish is made. Both exact quantities and timings are noticeably absent.

By the late sixteenth century, recipes start becoming slightly more expansive. *The Proper New Booke of Cookery*, published in 1575, includes a recipe for a 'Tart of Cheese' that specifies a soaking time of 3 hours for hard cheese. A decade later, *The Gude Huswifes Jewell* also features timings in some of its recipes, such as the one for 'a tarte of apples and orange pilles', which includes soaking times for the oranges in first water, then syrup 'for a day and a night'. Over the centuries, cookbooks become noticeably more detailed. Robert May, author of *The Accomplisht Cook* (1660), includes weights and measures in his recipes for dishes such as cheesecakes, also specifying details regarding ingredients, such as 'morning milk cheese curd fresh made'. 1747 saw the publication of Hannah Glasse's best-selling cookbook, *The Art of Cookery Made Plain and Easy*. Aimed at teaching servants how to cook, Glasse deliberately avoided, she tells us, 'the high, polite style' when she wrote it, opting instead for forthright, clear instructions. Some of her recipes do feature precise cooking times, such as a quarter of an hour to stew spinach, or half an hour to boil a custard pudding. In most, however, cooking timings are not given. In these, such as the one for fried lampreys, she simply writes 'when you think they are [*sic*] enough take them out', leaving the reader to use their own judgement as to the dish's readiness.

It is Eliza Acton, with her popular *Modern Cookery for Private Families* (1845), who brings a new structure and precision to the craft of recipe writing. Acton's recipes were much admired by no less a figure than Elizabeth David, who appreciated Acton's genuine depth of knowledge of food and cooking and her ability to share it concisely and elegantly. In her cookbook, Acton not only guides her readers through the stages of a recipe, but also concisely outlines the ingredients needed and gives overall cooking times. Her bread pudding recipe, for example, is summarized accordingly after the written method: 'New milk, 1 pint; sugar, 3oz; salt, few grains; bread-crumbs, ½ lb; eggs, 4 (5 if very small); nutmeg or lemon rind at pleasure: 1 hour and 10 minutes.' One of the reasons why nineteenth-century recipes featured timings to a much greater extent was the rise of the cast-iron closed stove. Offering a more uniform and consistent heat source, it allowed for timings to be usefully estimated.

Precise timings are nowadays taken for granted as part of the useful instruction contained in a recipe. It has become a convention to add timings for every stage of the process. These are intended by the recipe writer as a useful guide, but, of course, the danger is that the timings become too important, overly relied on. They are the kitchen equivalent of GPS for drivers, a friend remarked; one follows the instructions rigidly, without actually understanding where one is going. Timings in cooking depend on a huge range of factors – whether or not you preheated the pan and to what heat, how efficient your cookware is at conducting heat, whether the meat was fridge-cold or not, to name but a few – so should not be followed blindly. The judgements I make as a cook are to do with my physical senses; smell, sight, sound, touch all play their part. Looking at the way a sheen of oil ripples in the pan in order to judge how hot it is, listening to the sizzle as I fry, sniffing the fragrance of whole spices being dry-fried. Cooking involves an alertness to the reality of what is taking place in front of me, rather than following a recipe by rote and ignoring the evidence of my own eyes. Good recipe writing, in fact, uses timings *and* other information to explain to the cook what's being looked for. So an instruction for cooking rice might include the line 'Cover the pan, reduce the heat and simmer for 10–15 minutes' but will also go on to say 'until all the water has been absorbed and the rice is tender, dry and fluffy'.

Not only are recipes expected to contain timings, but the overall speed of recipes has itself become a selling point. Words and phrases such as 'fast', 'quick and easy', 'express', '15-minute meals', '20-minute meals' appear on cookbook titles, in food magazines and online websites. There is a hectic quality to titles such as Mark Bittman's *How to Cook Everything Fast*; on preparing meat, Bittman advises, 'Or cut into chunks. The smaller the pieces the faster it cooks.' Nigel Slater's seminal 1992 cookbook, *Real Fast Food*, with its typically intelligent approach to how to cook well but quickly, seems positively leisurely nowadays, allowing as it does a luxurious 30 minutes for preparing a 'real' meal. The most stylish of speedy cookbooks, however, appeared in France in the year 1930: Edouard de Pomiane's *Cooking in Ten Minutes.* A physicist at the Institut Pasteur in Paris, Pomiane applied logic and scientific knowledge of cooking techniques to his recipes. Thankfully, he also used wit and panache to convey his message. The advice in the recipes is simple but effective, such as not adding food to a frying pan until the fat is smoking hot, or cooking the flour for a béchamel sauce properly. He naturally appreciates the speedy potential of ingredients such as eggs and fish, observing, 'In ten minutes you have just time to boil a small fish. You have plenty of time to cook it in a frying pan, and still more time to deep-fry it in fat.' Among his 10-minute recipes are deftly executed versions of French classics such as *entrecote béarnaise*, calf's kidney with mustard, an omelette and *ceps à la Bordelaise.* In his preface, Pomiane offers a defence of his speed-centric approach to cooking, declaring that it is for 'everyone who has only an hour for lunch or dinner and yet wants half an hour of peace to watch the smoke of a cigarette whilst they sip a cup of coffee which has not even time to get cold'. The idea of cooking a rapid meal in order to linger at the table over a cup of coffee has a certain paradoxical charm to it.

THE QUEST FOR SETTING POINT

It is an annual ritual that I enjoy maintaining. Each February I buy small, subtly scented, baggy-textured Seville oranges, fetch out my large preserving pan and make a small batch of marmalade for home consumption. There is something pleasing about the process of making marmalade: small, patient, time-consuming steps, then the addictive excitement of the make-or-break finale.

First I halve the oranges and juice them. Next I tug out the inner membranes, pulling them away from the thick orange skins, an oddly satisfying task. These slippery membranes, with their pectin stickiness, are chopped. I wrap the pieces together with the seeds in muslin, to provide the pectin needed for my marmalade to set. I slice each orange skin into quarters, then sit patiently chopping the tough skin into fine slivers, with the knife releasing tiny spritzes of fragrance as it cuts into the peel. Once this is done, the chopped peel and parcel of pulp and pips are soaked in water with the juice overnight – to help soften the peel. Next morning the covered pan is set to simmer for 2 hours, filling the house with the distinctive bitter aroma of Seville oranges.

Once the tough peel has cooked through and softened, the drama begins, with just a few minutes in which to consolidate all the slow, careful work so far. I add sugar, stirring until it has thoroughly dissolved, then bring the mixture to the boil. Bubbles begin to appear, then bunch together, thickening to create a foamy mass. I begin testing for setting point by dipping a spoon into the mixture and looking at the way it forms on the spoon. There is a tension to this stage. I know from past experience that if I stop cooking too soon, my marmalade will be runny. If I leave it too long, I risk an overly firm marmalade, which is not to my taste. From the look of the mixture, I feel the right moment has been reached, turn off the gas and remove the marmalade pan from direct heat. There's a subtle glassiness to the texture of the mixture which looks hopeful. Having first let the marmalade rest for 10 minutes, skimming off any scum, I ladle it into warm jars – the pale honey colour of the jelly with shreds of dark orange peel suspended in it making me think of flies in amber – seal them and set them aside. As the marmalade cools, it thickens. Not being an experienced maker, the test comes when I open the first jar of that batch. What I want – and hope to find – is a jelly with a soft-ish set, a marmalade which instead of being over-sweet and caramelized has a bright zing of Seville bitterness. The sense of satisfaction when I achieve this 'rightness' adds to the pleasure of eating it.

Taking fresh fruit and cooking it with sugar to preserve it is something humans have done for hundreds of years. The way in which sugar preserves fruit is by creating a dehydrated environment that inhibits microbial growth, drawing out the water which microbes need to survive and grow. To make jam, two other elements in addition to fruit and sugar are needed, advises Pam Corbin, a noted jam-maker and widely respected judge at the World Marmalade Awards in Cumbria: 'acid and pectin'. It is using these in combination which make jams set. The acid (a solution with a pH of less than 7) commonly used in jam-making is citric acid, often in the form of fresh lemon juice. Pectin, a starch that gives structure to cell walls in fruit and vegetables, can be found in high levels in certain fruits. Blackcurrants, for example, are naturally high in pectin, making it relatively straightforward when turning them into jam to achieve a set. Other fruits, however, lack pectin, among them strawberries, which, despite being so popular, are famously a tricky fruit to make a jam from. Jam sugar, with added pectin, is, Corbin feels, a useful resource. 'If it means a novice can pick strawberries and make their own jam and know it will set, I think that's a good thing. Gooseberries have both pectin and acidity so are a great fruit to make a jam with. If you want to aid a strawberry jam in setting, put in a few gooseberries too. The important thing is to have enough sugar in your cupboard so when you've got fresh fruit you can get on with it and make the jam quickly. You want to preserve that fresh fruit as quickly as you can.'

Sky Cracknell and Kai Knutsen began their preserve-making business, England Preserves, in London in 2001, working initially out of their own kitchen. Both had grown up in families where jam was made, and when they wanted to set up a food business together, making jam presented an affordable opportunity, requiring little initial investment. 'We used to make jam at home that was very low-sugar but very runny, which wasn't commercially viable, so our holy grail for a number of years was to find a way to make a low-sugar, set jam,' explains Cracknell. 'The reason we made low sugar jam at home was that the taste was so much better – you get the taste of the fruit. The thing about sugar is that it can enhance flavour, but when you add too much sugar – pretty much all the commercial jams are at 63% sugar – you get very little distinction between the flavours, I find, especially with the more delicate flavours. Our jams have a 40% sugar content.' Within minutes of talking to her at their workplace in south-east London, Cracknell has put her finger on what it was that brought me here to see how they make jam. What is striking about England Preserves' jams is the depth of flavour of fruit they possess. I'm not their only admirer. While they started out selling to the public at Swiss Cottage farmers' market, their business today consists predominantly of selling to the food service industry, with the acclaimed French bakery Poilane among their customers.

The large England Preserves workshop in Bermondsey is at once relaxed and quietly efficient. There is an orderly air to the kitchen area, a sense of anticipation, with large jars and lids set out in readiness. From their early days, the process of jam-making has changed, as they learned and grew, with much trial and error along the way. 'It took time to improve our jam,' says Cracknell candidly. 'We do feel it has got better over the years. We've never rested on our laurels; even now we tweak the recipes.' In the kitchen area, Cracknell proudly shows me 'the most important bit of kit': two 32-gallon steam-jacketed kettles. These are gleaming stainless steel vats, around 5 feet high and 3.2 feet in diameter. Inside each of them is a large rotating paddle, enabling the kettle's contents to be stirred as well as heated. 'We invested in one of these about 10 years ago – a big step for us – and it really changed our lives.' Up until then, the couple made their preserves by hand. 'We worked so hard, all the time. I was always stirring pots of jam on the range all the time to stop it catching or burning. I had callouses for years on my hands from holding the jam spoon!' she laughs. The steam-jacketed kettle, as its name makes clear, has two layers, with steam injected in between. 'It heats up the sides as well as the base – very even – which means that you can raise the temperature incredibly quickly. Why that is important for us is because the quicker we cook the jam the better the flavour and colour.' 'The whole process is about speed,' adds Knutsen, 'to keep the brightness to the jam.'

The way in which England Preserves makes their jams begins

at the fruit farms, with whom they have worked for several years. The fruit is picked – usually slightly underripe as they find this better for jam-making – graded and then frozen at the farm; it is IPF, Individual Picked Frozen, rather than in blocks, a quicker process which retains the texture better. The fruit is then brought to London and cooked from frozen at England Preserves, as defrosting before cooking would mean losing the colour and the texture of the fruit. Their jams, it turns out, involve two ways of fighting the effects of time on fresh fruit: first freezing, then cooking with sugar. Once in the steam-jacketed kettle, the jams are quickly cooked, usually, depending on batch size, within half an hour. 'We do a high turnover of lots of little batches in our two kettles,' explains Cracknell. 'A big company would use 132-gallon kettles and have twenty-four of them!'

I watch a batch of strawberry jam being made, from start to finish a process spanning 30 minutes. 44 lb of frozen strawberries, varying in size, are tipped, with a splendid rattling sound, into one of the steam kettles, and lemon juice, for acidity, is added. The lid is closed and the cooking begins, with the paddle inside set to turn the fruit gently in one direction only. After around 10 minutes the lid is opened by Meghan Springer, one of the England Preserves jam-makers, to check how the fruit is cooking, releasing an appealing waft of strawberries into the air. Strawberry jam, it turns out, even for these experienced producers, requires diligence. They always cook it in these small batches, with each cook varying according to factors such as the size, water content and ripeness of the berries. Five minutes later the lid is opened again and some strawberries are fished out on a long wooden paddle. 'I'm touching the cores,' explains Springer, feeling the strawberries on the paddle carefully, 'as these are the hardest part of the strawberry. These aren't quite there yet.'

In the meantime, a second batch of strawberry jam is set off in the second kettle. What is striking to me, watching the process, is the way in which, without the use of timers to prompt, Springer checks the cooking at key stages. 'I check what the time is at each stage and keep an eye in my head on the process. We say a lot of things out loud while we work, like "I've just put in the strawberries!" It is tricky when I'm doing two different jams.' The strawberries are looked at and touched again to gauge readiness. 'Of course, some of the fruit cooks down,' explains Cracknell, 'but we want to keep some texture, have pieces of fruit too. But the fruit must be cooked through, so you want some strawberries to be whole enough to scoop out of the jar, yet soft enough to break down on your knife when you spread it.' It is important to get the cook right at this stage, as this will affect the colour and the texture of the jam.

With the strawberries judged to be the right texture now, pectin (mixed with a little of the total sugar to give it weight and make it mix better) is added to them. A few minutes later, the set is checked. With the addition of the pectin, I can see, glancing into the kettle, that the jam now has a glossiness and a more vibrant colour. Springer stirs the mixture and takes out a small amount of jam with the wooden jam spoon, depositing it on the metal side of the kettle. She and Cracknell glance at it with satisfaction. 'You can see the shape is rounded,' explains Cracknell. 'It has an inherent viscosity and is holding on to the surface.' A few minutes later, when it's cooler, Springer checks the jam sample, pushing it gently across the surface. 'It's moving as a mass – that's a good sign.' The time has come now to add the majority of the sugar. This is stirred through and the mixture is brought to temperature – around 208°F. In the meantime, large sterilized 3lb 14oz jars and lids have been put out, ready for potting.

A few minutes later the jam is ready and a sense of purpose is apparent in both Springer and Cracknell. The rich, scarlet, textured liquid with strawberries in it is swiftly scooped out and poured into jars, sealed with lids, each jar then turned over to ensure a heat seal. This stage has to be done quickly for reasons of hygiene. Potting hot strawberry jam into large jars, however, brings with it an issue, explains Cracknell ruefully, as the fruit rises to the surface as it cools. 'After about 2 hours of cooling, we shake each jar of jam to disperse the fruit evenly,' she explains, to my surprise. 'Yes, a good shake, not a gentle one!'

There are commercial implications to working in the way England Preserves does. While using a low sugar content brings results in terms of flavour and texture, it restricts the shelf-life of their jams as the colour of their lighter jams fades noticeably. One of the things that a high sugar content in jam does is retain colour. 'We don't carry much stock,' says Knutsen, gesturing to a largely empty stack of metal shelves in the storage area. 'We're practically making to order. It's not unusual for a shop to be stocking one of our jams made 24 hours earlier. We're basically balancing along the edge of possibility because of the limited shelf-life' – a glint in his eye.

The results of producing jams as they do – cooked quickly in small batches – is clear when I look at the England Preserves range. That commitment to capturing fresh fruit in jam form is apparent visually: bright red strawberry jam, deep orange Bergeron apricot jam, dark red raspberry jam, purple damson jam and – leaping out – the fluorescent pink of Yorkshire forced rhubarb jam. The flavours dance on my tongue – from a noticeably tart blackcurrant to a tiny jar of wild strawberry, which tastes of violets. These are jams for people who love fruit. The texture is very far from the jellied solidity of a mass-produced jam, made to a rigid, repeated formula. While the England Preserves jams are set, there is a softness to their texture, the jams lending themselves to stirring into yoghurt or porridge, as well as being spread on a slice of bread.

Having seen Cracknell and her team at work, I realize that while a process was followed, it was not done so rigidly. What was important was looking at each batch of fruit, noticing how it was cooking – tasting, smelling and feeling the fruit were vital. Human judgement, rather than a prescribed formula, determined when setting point had been reached.

Encouragingly, for domestic jam-makers, cooking quickly in small batches is very achievable. 'When people are making jam at home we always advise making small batches,' says Cracknell emphatically. 'I think the mistake is that people buy a big preserving pan, fill it up to the brim with fruit and try to make this massive amount of jam. That will be bubbling away for hours, by which time it will lose all the colour and flavour, turning into syrup.'

Thinking of the care and attention that went into making the jams at England Preserves, I'm reminded of something that preserving expert Pam Corbin said: 'We talk about four ingredients – fruit, sugar, acid and pectin – you also need a bit of love as well.'

HOURS

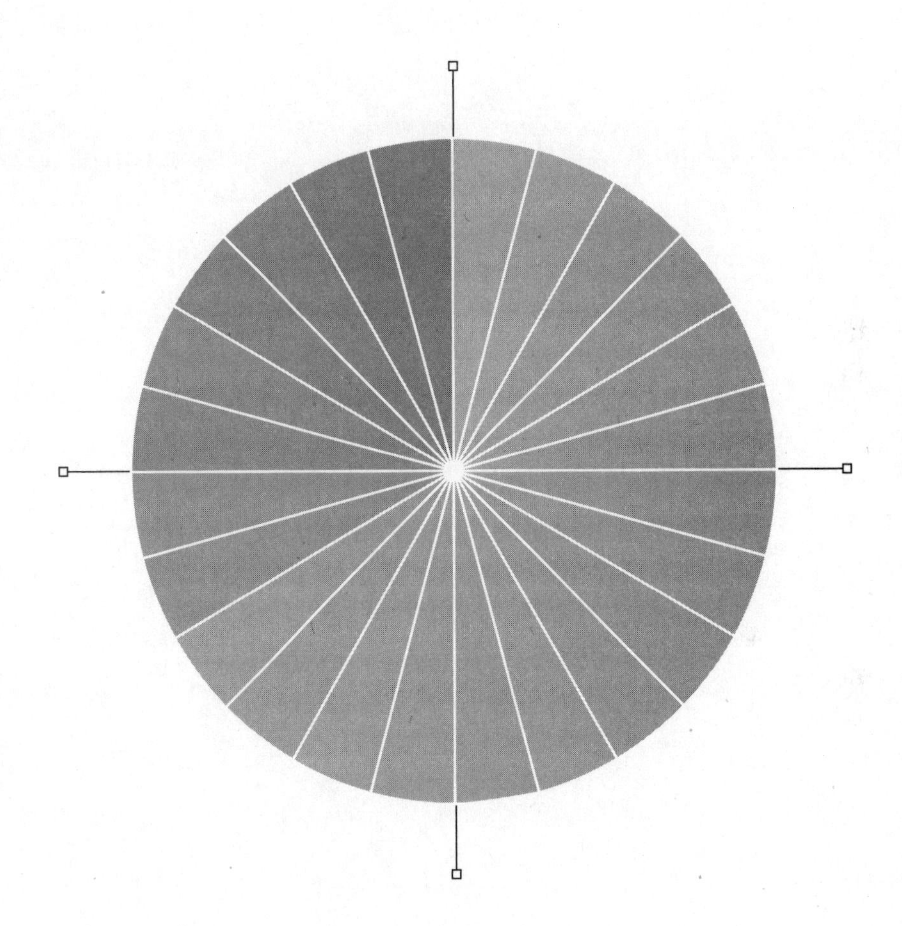

HOURS

HOURS

In contrast to the precision often required with minutes, there is a forgiving quality to hours. Many of the procedures that take hours – marinating, braising or gentle roasting at a low temperature, soaking pulses – have a flexible aspect, allowing for a range of hours to be used. Traditionally, breads – both sourdough and yeast – took hours to rise, until the invention of the Chorleywood method in 1961 speeded up the whole process enormously. The renaissance of craft bakers in the UK over the last two decades has seen time put back into the making of bread, as in other processes. As time has been squeezed out of food and cooking, its ample use has acquired a new value. There is now an appreciation of time taken in cooking, with slow-cooked dishes enjoying a cult appeal among both home cooks and restaurant chefs.

THE CRAFT OF BREAD

Hot buttered toast in the morning for breakfast, a convenient sandwich for lunch at work in the office, a slice of bread with supper at night – bread is one of our staple foods. Simple and basic, it has nourished human beings for thousands of years. Flatbreads are its earliest form, eaten since the Stone Age. Leavened (raised) bread, while a more recent development, has also been eaten for millennia, with archaeological evidence for its existence found dating back to Ancient Egypt. For several centuries, yeast was added to dough in the form of a sourdough starter, made by using wild yeasts in the air. The way in which yeast – that extraordinary, resilient fungus – works to raise dough is by consuming the sugars in the flour and producing CO_2 as a by-product. This CO_2 is then trapped by a stretchy, elastic net of gluten (made by mixing flour with liquid), creating tiny air pockets, causing the dough to swell and rise and giving yeast-leavened bread its distinctive spongy texture. This biological rising process, triggered by a sourdough starter, takes hours to happen. During the nineteenth century the work of Pasteur saw a better understanding of yeast develop and the commercial production of what is known as baker's yeast, *Saccharomyces cerevisiae*, a powerful, effective raising agent. During the twentieth century, bakers in Britain began to prefer it to the traditional, slower-working sourdough starters made using wild yeasts. A thrifty way in which bakers worked with this expensive baker's yeast was to add just a small amount of it in a preliminary 'sponge' made from flour and water and set this aside to ferment, allowing the baker's yeast to multiply and so go further.

It was, however, in 1961, in a leafy corner of south-east

England, that a huge, revolutionary step in the history of bread-making took place. Scientists at the British Baking Industries Research Association laboratories at Chorleywood, Hertfordshire, came up with the Chorleywood Bread Process (CBP), a mechanized way of making bread that considerably speeded things up. Key to it is the powerful, rapid formation of the dough. Whereas traditional kneading by hand might take around 20–25 minutes, in the Chorleywood Bread Process the ingredients forming the dough are violently agitated for around just 3 minutes, a process requiring a large amount of energy and heat. In order for this method to be successful, however, a number of things have to be added to the dough mix. First, it has to have a higher amount of baker's yeast than had traditionally been used. Further essential additions include hard fat to help create a soft crumb; emulsifiers, which enable the dough to hold more gas; and enzyme-based 'improvers', the latter classified as 'processing aids', which means they do not have to be declared on product labels. Using the Chorleywood Bread Process means that the journey from flour to baked loaf takes just 2–2½ hours, plus an hour for cooling. The bread itself is bland, soft and light, with a candyfloss insubstantiality to it. During eating, in contrast, this fluffy bread sticks to the roof of one's mouth.

Hailed as a way of allowing bakers to make affordable bread quickly and easily from low-gluten British wheat, over 80% of bread in the UK is now made using this method and its usage has spread around the world. What is striking if one keeps a loaf of CBP bread for a few days is that it does not dry out and go stale in the way traditional bread used to, resulting in frugal recipes to use it up, such as panzanella – instead it simply begins to go mouldy. One school of thought links the rise in gluten intolerance to this high-speed way of manufacturing bread. In his magisterial tome *Bread Matters*, baker Andrew Whitley writes of modern industrial baking: 'Time was squeezed out of the baking process, and with it not just flavour, but vital nutritional benefits.'

Britain in recent decades has seen a revival of craft baking across the country, focused on using traditional methods rather than the Chorleywood Bread Process. In the picturesque Suffolk market town of Woodbridge, the Cake Shop Bakery occupies a special place in the community, a family-run establishment and the town's oldest business. Situated in a prime position on the Thoroughfare, the town's main shopping street, the bakery and shop occupy a spacious site, housed in Victorian red-brick premises, its presence marked with a string of Union Jacks beneath a large sign. Inside, the display of breads, cakes and biscuits is extensive and noticeably varied. The stock ranges from traditional items rarely glimpsed in craft bakeries, such as English muffins and tea cakes, to focaccia, 72-hour sourdough and loaves of Polish rye. The popularity of the bakery is witnessed by a steady stream of customers; by the time I leave, there is a queue trailing out of the door, waiting patiently to be served. Founded in 1946 by Jonty Wright, the fact that this bakery has endured, when so many bakeries have vanished from our high streets, is a tribute to one family's commitment to keeping their baking tradition going.

Baker David Wright, together with his four sisters, took over the running of the family bakery founded by their grandad in 2012, when their parents, after years of hard work, decided to step back from the business. A tall, thoughtful, softly spoken man in his early thirties, David had been working as an actor and director in London, although often supplementing his income by baking in commercial bakeries in the capital as well, due to his experience of the business while growing up. Wright explains to me that the way he and his siblings approached breathing new life into the bakery was founded upon a respect among this new generation for the legacy of the past. 'One of the key things for us taking on the bakery,' Wright explains, 'as it's been around for so long and people have treasured memories of it, was not to come in and change everything. Any changes we did make were grounded in a historical nod, so looking back at what my grandma and grandad did, what Mum and Dad did, what maybe had been successful but fallen out of favour over the years, which we could then resurrect and add our own spin on. A way of doing something new, while still staying true to the history of the business.'

The sheer amount of hard work that goes into making their array of breads and baked goods becomes clear as I talk to Wright. When I turn up at the shop to meet him at midday, he has already been long at work, beginning his shift at 1.30 a.m. As sourdough bread famously takes several hours to prove, this long, slow period means that it can be prepared in advance, set aside to prove, then baked, making it a relatively flexible part of the baking process. In contrast, the traditionally made yeast bread using baker's yeast has a much shorter window of making, and it is the preparation of this which requires the very early morning shift. 'The way we make our white bread, for example, is pretty much the way that Grandad made his white bread, which was the way that bakers made it for a long time before: the sponge method,' explains Wright. 'We make a preferment – the sponge – and then add to it the morning that we're baking, so our yeast bread has that fermented element to it. It gives it more flavour, and one of the reasons that bakers would use it is that the time you can use it for is elongated. Because its development is slower, the window you can cut and shape the dough is longer. If you take a massive amount of dough, say over 220lb of dough, and you're hand-processing every single loaf, which is what we do, then it takes some time. With a quick yeast dough, by the time we'd finished processing it, it would be spent. The yeast would feed on all the sugars and you wouldn't get the same results.' Throughout the history of the bakery, Wright's family has never used the Chorleywood process because, says Wright simply, 'We're craft bakers.' Although it would make life easier– and do away with the punishing early-morning shifts – it is not how they approach baking. It turns out that the hours required by working traditionally involves the small Cake Shop Bakery employing six to seven bakers, enough to staff five or six large plant bakeries.

The community aspect of the Cake Shop Bakery in meaningfully serving the people of Woodbridge is important to Wright and his family. 'The main reason we've survived is that people round here have continued to support us. Even through the advent of supermarket [Chorleywood Bread Process] bread, we've managed to keep our customers and I think the main reason we've done that is through diversifying and by keeping the quality up as well. Small bakers tried to imitate supermarket bakeries by bringing time down. We feel the way of competing with industrial bakeries is to do what they can't do. So while their object is to shorten the amount of time the process takes to maximize profitability, we went down the route of making breads which could only really be made by using a long amount of time.' This requires commitment, I observe. 'When you're in a family business and members of your family are in the workforce you can exploit them!' laughs Wright. 'I think if we sat down and worked it out we're all earning less than the minimum wage, but that's what you get with a family business.'

While maintaining the craft and skill of traditional baking is key, so, too, admirably, is an emphasis on accessibility. 'If you're making good bread and selling it at a really high price, all you're really doing is serving the supermarkets who are producing low-quality, low-price bread,' points out Wright. 'You're pricing quality bread out of the market and pretty much forcing people to that supermarket bread. The way to make sure that people have an alternative to that supermarket mass-produced bread is to produce something that is affordable and accessible. We definitely feel that the bread that we make is the kind of bread we can sell to everybody.' Given this philosophy, I am unsurprised to hear Wright declare, 'The best-selling thing that we do is a white sandwich loaf and, to be honest with you, I feel that's a feather in our cap because we're making this good white sandwich loaf which is competing with industrially made bread and people are still choosing ours. It requires a lot of skill to bake in a tin because you have to achieve the right shape within it. When you bake freeform you're looking for these wonderful organic shapes, whereas with the tin if you don't let it prove to exactly the right level, either it will prove too much or it will prove too little, and either of these things means that you have ruined the loaf.'

Wright shows me around the family bakery space behind the shop; housed in a Victorian building, it is spread over two levels. We pass huge sacks of flour, stacks of old stainless steel bread tins – square sandwich tins, farmhouse tins with rounded corners – and provers, resembling metal cabinets, in which the dough is left to prove in warm, moist conditions. There are dough-dividers, which cut dough into even portions, and a variety of mixers. Much of the machinery is vintage, I note, learning that it has been passed on through the generations. 'I've had to repair a lot of kit,' he tells me. He pats the huge oven with palpable affection. 'We spent a lot of money repairing this. It's a 1980 Werner & Pfleiderer deck oven – they're like the Rolls-Royce of ovens, so well built.' He opens one of the five decks to show me how deep it is: each deck can hold forty trays, with each tray containing seven tins. It is, Wright explains, a piece of equipment with idiosyncrasies and there has been a process of learning how to use it well. Although using the oven efficiently is important, if it's filled 100% there won't be a good bake as it won't all cook evenly, so it is important to keep some space in it. Certain products bake best in certain parts of it, with, for example, the French Country loaves baking best in the top deck. 'You'll notice there's no timer and that you can't see into the oven. As you can't keep opening the oven during baking, one of the main skills of being on the oven is that you need to know exactly what's in the oven and where it is without being able to see it. In your head, you need to know exactly what is going in next and where; you're always planning ahead.'

Each day's routine begins with switching on the oven. The sourdough loaves, shaped the day before and refrigerated, are the first breads to be baked. In the meantime, the yeast bread doughs are prepared, proved and shaped. Limited space on the tables on which the dough is moulded, the provers in which it proves and the ovens means careful organization is required. There is a sense of an orderly pattern to the process David describes. 'You get very smooth natural rhythms in processing the dough, proving, shaping and baking, especially when you've been working with people for a number of years and the production is very similar from day to day in terms of what you do and in what order. The person mixing the dough has to be aware of everything going on around them, as does everyone else, without talking about it. There's a kind of connected rhythm.'

I step out from the bakery area into the shop, filled with the results of David's and his colleagues' labours. The steady stream of shoppers means that the stock of bread, cakes and biscuits is diminishing before my eyes. Wright, by now palpably weary, says goodbye to me and heads off to go to sleep. I join the queue and stock up with freshly baked bread, including, naturally, a white sandwich tin, and a couple of snacks to take back to London. When a few hours later I arrive home, I put on the kettle and make a cup of tea. One of the snacks I've bought is the appealingly named Devon split: a simple white roll, sliced open and filled with buttercream and topped with colourful sprinkles. Eating it makes me feel like a child again. The plainness of the bread contrasts with the sweetness of the filling. It is, I realize, a creation that relies entirely on the quality of the bread to succeed in being pleasurable.

RITUAL ROASTING

The roast meal, often called simply a 'roast', has a special place in British culture. There is the idea of the weekly 'Sunday roast', when families gather round a table to share a meal. British Christian festivals are traditionally celebrated with roast meat – roast lamb at Easter and roast turkey at Christmas. The British love of roast beef was celebrated in Henry Fielding's eighteenth-century patriotic ballad, 'The Roast Beef of Old England', with the French nicknaming us *les rosbifs*. Roast meat, of course, carries a cachet in many other cultures around the world, from Thanksgiving turkey to the roast suckling pig served at Chinese wedding banquets. It is a meal associated with hospitality and plenty.

Our sense that a roast meal is something special goes back far into our history, as Marc Meltonville, food historian for the Historic Royal Palaces, makes clear when he meets me at Hampton Court Palace, East Molesey, Surrey, with its magnificent, huge Tudor kitchens. 'Why did they dedicate a room the size of two tennis courts simply to roasting with six fires?' It is, he tells me, to do with maintaining social status. 'Fresh meat is an expensive luxury. You then use the most fuel-inefficient way of cooking, which is roasting, which gives a beautiful, flavourful result. You've got to pay a team of people to stoke the fire and to turn the spit.' Roasting, in culinary vocabulary, is a specific term, meaning cooking using radiated heat from one side, on a spit. 'Over a fire is grilling, to one side of a fire is roasting, and in a box, like an oven, is baking.' With the rise of ranges, however, Meltonville points out that even though we bake bread, biscuits and cakes, we 'roast' a chicken. 'No, you don't, you bake it. But we flick that word round to elevate the meal. Roast has remained, even in our twenty-first-century psyche, a good, special thing.'

The logistical realities of producing a roast meal, even in today's ovens rather than using an open fire, are presented as challenging. Serious-minded cookbooks offer tables of recommended roasting times, for meats such as beef, pork, lamb, poultry and game birds, based on a ratio of weight to height to time. Beef and lamb bring with them the question of how well cooked one wants them to be: rare, medium or well done, affecting the roasting time accordingly. Poultry, of course, for reasons of food safety, *must* be cooked through properly, a cause of anxiety for any conscientious cook, especially when faced with a large, heavy turkey. A further complication is that a 'proper' roast meal also requires trimmings, such as roast potatoes, parsnips and stuffing, or, in the case of roast beef, Yorkshire pudding, which also need to be cooked in the oven. The tricky part is that often these side dishes require to be cooked at a different temperature from the main centrepiece. This makes co-ordinating challenging, as anyone who has struggled to put all the elements of a traditional Christmas meal on the table at the same time will testify. A roast meal is all about timings, it turns out. When it comes to festive roast meals such as Christmas lunch or Thanksgiving dinner, much column space each year in the pages of newspapers and food magazines is devoted to how to cook them. In these stressful times, reassuring, culinary gurus such as Delia Smith and Martha Stewart are turned to for support. Detailed schedules with step-by-step timings are offered for guidance, giving the sense that the event is a military campaign rather than a celebration. The cook's countdown for Christmas Day often begins early in the morning with an instruction to preheat the oven. While an understanding of how long to roast meat or poultry for is required, one kitchen gadget, however, has removed a lot of uncertainty from the process: the probe thermometer.

THE BEAUTY OF THE BAIN-MARIE

The bain-marie comes into its own when slow, gentle cooking is required, for example when you want to coax temperature-sensitive dishes such as egg custards into solidifying and setting without overheating. Its origins are alchemical: *balneum Mariae* or Mary's Bath, named after Mary Hebraea (Mary the Jewess), a legendary alchemist thought to have lived between the first and third centuries CE and credited with inventing it. As the name implies, water is central to the concept of a traditional bain-marie. Warm water is poured in a protective layer around the dish that is being cooked, contained within another, larger vessel. An important factor in the process is that the water should not be allowed to boil during the cooking period. The water forms an insulating barrier between the direct heat and the dish, allowing for slow, gentle cooking.

The use of a bain-marie is central to France's terrines and pâtés. Using the correct amount of hot water is important to the successful cooking of these dishes. The hot liquid is carefully poured into a container so that it comes only halfway up the sides of the terrine or pâté dish, allowing enough heat to penetrate the mixture to cook it through. The cooking times, as Jane Grigson points out in her *Charcuterie and French Pork Cookery*, vary according to the depth of the dish: 'small but deep dishes of pâté take longer than wide shallow ones'. Using a bain-marie – in which the heat is gently transmuted through water – is an excellent way of retaining a moist, delicate texture in dishes such as cheesecakes. In Chinese cooking the use of a traditional soup pot – a double-lidded ceramic pot – follows the same principle. The pot is filled with soup ingredients, such as chicken bones, water and, often, medicinal herbs, covered, then placed in simmering water. Soups cooked like this are regarded as especially nutritious and health-giving.

The term 'bain-marie' is also used for another way of cooking with water, sometimes called the double boiler method. In a classic double boiler, hot water is contained in the bottom part, while the dish being cooked is suspended above, importantly not directly in contact with the hot water. In this method, too, the heat is transmitted through the water, permitting certain ingredients to be cooked successfully. When melting chocolate, for example, this type of bain-marie is called for, since if the chocolate scorches as it melts it will 'seize', becoming greasy and grainy. Zabaglione, a classic Italian dessert tracing its origins to seventeenth-century Turin, is one of the more dramatic creations made using a bain-marie or double boiler. To make it, first egg yolks and sugar are whisked together until pale and creamy, then Marsala is mixed in. The mixture is placed over a bain-marie of simmering water and whisked continuously, swelling to form a thick, pale, foamy mass. It is a process that takes around 15–20 minutes, so patience at the stove is required. The results – a warm, boozy cloud – must be eaten at once, which is no hardship at all.

POTS AND PANS

One of life's small pleasures for someone who, like me, enjoys cooking is using the 'right' piece of kitchenware for the task at hand. A phrase I often use in my own recipe writing is ' heavy-bottomed pan'. Classically, I would stipulate this in a recipe for a slow-cooked dish, such as a casserole or a custard, or for the cooking of rice by the absorption method, with the heavy base allowing for heat to be conducted gently and evenly over a period of time as required. Trying to get the same results using a thin, flimsy pan would be harder, given the risk of scorching or curdling or boiling dry. Heavy-based pans are also excellent heat conductors, so other, speedier, aspects of cooking can take place. Browning meat, for example, to create the desirable Maillard reaction, can be done swiftly and well in a thick-based frying pan, achieving browning rather than burning. The solid, weighty cast-iron griddle pan that I lugged back from Italy is far more effective at griddling food than a thinner, lighter, non-stick aluminium pan.

Over the centuries, much human ingenuity has gone into finding materials that work well in the kitchen. Certain types of cookware are particularly appropriate for slow cooking. Morocco's distinctive lidded, conical tagine dishes, made from clay, are ideal both in form and material for the gentle melding of meat, vegetables, spices and water that creates the eponymous stews, traditionally cooked for hours in bakers' ovens. As the steam rises within the dish, it re-condenses at the top of the lid, falling back down as liquid, so preventing the tagine from drying out as it cooks. Earthenware pots have been used for slow-cooked dishes for centuries, from Colombia's traditional clay pots to China's sandpots, with their distinctive rough texture. In contrast, China's wok is ideal for the speedy, vigorous action of stir-frying. Its small base concentrates the heat effectively, while the wide, sloping sides give enough space for the ingredients to be moved within it. The making of marmalade begins with me hunting out my preserving pan, a large, wide pan that enables the mixture to reduce quickly and efficiently without cooking out the flavour of the Sevilles.

Keen cooks become attached to their kitchen kit, with long-lasting trusty items taking on a companionable quality over years of use. British cookery writer and broadcaster Thane Prince bought a large orange Le Creuset casserole dish nearly forty years ago in a sale at Heal's. 'I bought it for lust,' she says simply. 'I looked at it and wanted it. I think big cooking pots are about expansive food, cooking for crowds. When I saw the pan, I saw people sitting and eating.' When her children were young, she used it for making batches of chilli con carne or Bolognese sauce with which to stock up her freezer, with jam-making its primary use nowadays. As she points out, the advantage of a heavy-based pan is the range of one-pot cooking it permits, from slow sweating of onions over a low heat 'until softened and mellowed' to increasing the heat to brown meat or chicken before adding stock, covering and cooking in the oven. 'Having the right pan to cook with just makes the whole process much easier,' she asserts. 'Both my daughters have Le Creuset pans given to them by me; I felt it was important. No doubt they'll be after my pans, when I go.'

When I want to buy a piece of kitchenware, either for myself or as a present for a friend, it is to David Mellor in Sloane Square that I go. The shop, opened in 1969 by the British designer David Mellor, is a discreet, elegant presence. Here one finds the famous cutlery, made to this day in the Peak District by the company he founded, as well as tableware and kitchenware. There is a pared-down, harmonious quality to the displays: a row of thick, heavy glass bowls in varying shades of blue and green, a line of knives and forks, a neat pile of wooden chopping boards. Over the decades I've noticed a great continuity to their stock: the same brightly coloured Scandinavian glasses, stocked years before their current fashionability; rustic, brown-hued Leach pottery; gleaming copper pans by Mauviel. 'Really the last thing you want is fashion in the kitchen,' Corin Mellor, creative director, says emphatically. 'Particularly with tableware, I think people should be able to come back in twenty years' time and elongate their collection.' In stocking the shop, he explains, the focus is on longevity, things that will last and give good value. Many of the companies they buy from share similar values to David Mellor's, he reflects. A designer himself, Mellor is clear that the items they carry must both do their job well and look good. 'Normally form follows function, so if something works it looks good too,' he observes. His favourite material is stainless steel. 'You can do an awful lot with it. You can bend it, cast it, have different degrees of hardness. It's a very versatile material and is also pretty indestructible.' Knowing that what I buy from Mellor's will be well made – will last for years – has been one of the factors in my own customer loyalty. I am not alone in my fondness, as the business is thriving and they have opened a second London shop in Marylebone. The classic stock – well-wrought pans, hand-turned wooden salad bowls, French steel omelette pans – always makes me think of Elizabeth David. 'She did know the Sloane Square shop,' Mellor tells me. 'We carry her cookbooks. We're very much from that era. Perhaps it's come round again!'

PULSE OF LIFE

Drying food has long been a means of preserving and safeguarding it for future use, saving food harvested during a season of plenty for leaner times ahead. Pulses – the term used to describe the dried seeds of the bean and pea family – are historically an important dried food, an invaluable source of protein in countries such as India and Egypt. Step into an Indian food shop and you see at a glance how important pulses are to the Indian diet. Here are shelves lined with packets containing a wide array of whole and split dried pulses, sold at noticeably reasonable prices; the colour palette ranges from the ivory white of split, skinless urad dal through green mung beans to purple-black kidney beans. The culinary diversity of this valuable dried food is extraordinary, used to make everything from soups and dips to stews and pancakes. For centuries in Britain the word 'peas' – which today makes us think instantly of the small round bright green vegetable – was synonymous with dried peas, used in hearty, traditional dishes such as pease pudding or pottage. In British cuisine, however, pulses fell out of favour a long time ago. As we became a wealthy, industrialized country, the pulse became stigmatized as the food of the poor, which indeed it was. People replaced this cheap source of vegetable protein with protein from meat and dairy products, a move which has been reflected in every pulse-eating nation as it becomes more affluent. In today's age of quick fixes, the idea of the pulse – this humble, fundamental ingredient, which requires time to soak and cook – lacks appeal. The gentle pleasures of pulses – their subtle flavours and textures – are often overlooked, even by people who enjoy cooking and eating.

Nick Saltmarsh, a tall, thoughtful man who talks about pulses with a quiet but evident passion, is determined to change that. His company, Hodmedod, which he co-founded with Josiah Meldrum and William Hudson in 2012, was created to promote and sell British pulses, reviving our relationship with our native ones, formerly an essential part of our diet, now barely acknowledged. The starting point for Hodmedod's was a sustainability project in Norwich with the Transition Towns Group, looking into locally grown sources of vegetable protein. Saltmarsh and his colleagues realized with astonishment that fava beans are grown widely across Britain, but are simply not eaten here, exported instead to Egypt where *ful medames*, made from dried fava beans, is a national dish. 'We tried them and thought they were very good,' says Saltmarsh, 'and wanted to bring British beans back to the kitchen.' Britain's relationship with fava beans is, he tells me, one stretching back millennia. Fava were one of the very first crops to be introduced by Iron Age farmers. 'They would have been one of the main crops grown here, as a bean that was harvested dry which could be stored as a source of year-round protein. And, alongside dried peas, it would have been pretty much the main source of protein in the British diet.'

Fava beans, in contrast to New World beans such as kidney, borlotti and flageolet, introduced to Europe from the Americas, grow admirably in Britain. Being tolerant of frosts, they are able to be planted either in September/October or February/March and harvested in August. 'Beans and peas were historically almost entirely harvested dry. We're remarkably out of touch with the fact that all pulses dry on the plant,' Saltmarsh muses. 'They may need just a little more drying out. From harvest, you can store them for a year or even longer. They last almost indefinitely really, although the longer they dry, the harder they become to cook, though it's only after 3 years that that becomes noticeable.' Dried on the plant, straightforward to harvest and requiring little work to store, pulses, it seems, are a food where nature does a lot of the work. Furthermore, harvesting the pea or bean when it is dry means you are harvesting a food at a stage when it is very nutritious indeed. 'Around the time we stopped eating so many dried pulses, we began to eat the immature peas and beans as a fresh crop. Because they are immature a lot of the sugars inside them haven't gone through the process of converting into starches and the protein content hasn't developed as it has in the fully ripe pea or bean.'

With the emphasis on promoting the eating of mature peas and beans, the Hodmedod range has expanded to include such little-known British pulses as the Carlin pea, strikingly dark brown or reddish-brown in colour, and whole blue peas, which are the same peas used for split green peas but with the skin left on. We discuss the question of soaking pulses before cooking, often seen as a 'nuisance' when it comes to using them. Pulses, Saltmarsh tells me, come in either the whole form, with their skin on, or the split form, where the skin is taken off and the pulse, as a seed, naturally falls into two halves. 'All the split pulses can be cooked from dry without soaking.' Whole pulses, however, do require soaking, with a good rule of thumb, I am told, being 6 hours as a minimum period. While the process requires time, it is, of course, simple and unlaborious. 'If you soak them for a long time, they will start to sprout, because they are alive. They are seeds and they want to grow.' There is a tradition in some countries, he tells me, of allowing pulses to partially sprout before cooking them, this process improving the availability of some nutrients. 'We tend to think of pulses in very binary terms – you either cook them unsprouted or you sprout them and eat them raw.' Saltmarsh's fascination with pulses as a valuable source of protein – and, through their nitrogen-fixing properties, a crop that remarkably puts fertility into the soil rather than extracting it – is clear. 'When we started we had' – he pauses and laughs gently – 'and still have, the challenge of developing awareness and understanding of these British pulses, which as a country we've lost contact with. There is a resistance; a lot of people don't think of cooking with dried pulses. Once you've soaked them and cooked them a few times, however, and just got into the rhythm of what you need to do, it becomes very easy.'

MARINATING

Marinating is among those kitchen procedures that require the cook to think ahead and prepare in advance, the use of time being an intrinsic aspect of the process. Marinades, which range in texture from liquid mixtures to pastes, require contact time with the food in order to be effective, although this will vary depending on what is being marinated. This contact time brings a satisfyingly transformative element to ingredients. Through marinating, everyday foods – chicken legs, tofu, lamb chops – gain a new, vibrant flavour dimension.

There are two primary reasons for using marinades: to tenderize and to add flavour. The tenderizing imperative means that acidic ingredients, which act to break down cell walls, are a key component in marinades around the world: for example, green papaya (high in papain, an enzyme that weakens the protein bonds in meat) or lime juice in the tropics, yoghurt in Middle Eastern and Indian cuisine, buttermilk in the USA and red wine or vinegar in European cooking. Many traditional recipes for game meat – often tough, from older animals that have lived active lives – require an initial marinade for this purpose. Elizabeth Luard's recipe in *European Peasant Cookery* for a Spanish dish of rabbit with garlic involves skinning and jointing a rabbit, sprinkling the meat with 3 tablespoons of white wine vinegar and leaving it to marinate overnight. 'The marinade softens both flesh and flavour and is more necessary with a rabbit than other game,' she observes.

In order to create taste, the flavourings in marinades are gutsy, big hitters: garlic, root ginger, onions, herbs and spices. For maximum effect during contact time, these are often pounded together to release their flavour compounds. Creating a paste in this way also allows the ingredient being marinated to be thoroughly coated. The traditional South East Asian dish of satay, for example, features pieces of beef marinated in an aromatic paste made from lemon grass, shallots, garlic, galangal and spices, the latter dry-fried to increase their aroma. A marinade paste of fiery Scotch bonnet peppers with herbs and spices gives Jamaican jerk chicken its distinctive kick.

When it comes to how long to marinate, recipes vary enormously in suggested time span. As a logical rule of thumb, small pieces of meat or poultry require shorter marinating times than large cuts or whole birds. There is often a marked flexibility to suggested marinating times. Yotam Ottolenghi and Sami Tamimi's popular lamb shawarma recipe in their best-selling book *Jerusalem* suggests marinating the leg of lamb for either 2 hours or overnight.

It is, of course, possible to marinate ingredients for too long, over-saturating them and tenderizing them to the point that they lose texture. In his fascinating book *The Food Lab*, J. Kenji López-Alt advises that the best marinades contain a balanced mixture of oil, acid and salty liquid. Excessive acid, he explains, begins to chemically 'cook' the meat, denaturing the protein and causing it to become firm. His recommendation for acid-containing marinades is to use equal parts acid and oil 'and limit exposure to under 10 hours'. The risk of adversely affecting texture through over-long marinating is especially pronounced in the case of fish and seafood, due to their delicate cell structure. Madhur Jaffrey recommends a mere 30 minutes' marinating time for tandoori prawns, but 6–24 hours for tandoori chicken.

It is striking that one of the world's great marinated dishes, Peru's ceviche, is now prepared in minutes rather than hours. The venerable Central American tradition of making ceviche from fresh, raw fish, marinated through contact with an acidic fruit, dates back thousands of years. In Peru, *tumbu* (banana passionfruit) was historically used, Peruvian restaurateur and chef Martin Morales tells me, but the arrival with the Spanish conquistadors of citrus fruits such as limes and bitter oranges saw their juice take precedence. In Peru the making and consumption of fish ceviche was linked to the fishermen's working day, which began at dawn. When the boats returned in the morning, their catch of the day was transformed into ceviche and eaten for lunch, never for dinner. The traditional marinating time for ceviche in Peru in the 1970s was 2 or 3 hours: 'It was like poaching fish in lime juice,' Morales tells me, 'totally white.' Today, preparing ceviche has sped up, allowing the taste and texture of the fresh fish to be appreciated as well as flavours such as zesty lime, salt and piquant chilli. Morales is adamant that fresh ingredients and a freshly made marinade – the evocatively named *leche de tigre* (tiger's milk) – with a contact time of simply 2 minutes – guarantee optimal flavour. 'It's about creating excitement in your palate. If you lose the impact of the flavours, it's like arriving at a party when it's been going on too long – it's boring!'

TAKING STOCK

The making of stock is both straightforward and satisfying, a gentle process of extracting taste and goodness from ingredients such as bones, vegetables and herbs through cooking them in water. Over that period, the water is enriched and transformed, taking on the flavours of what has been simmered in it. As anyone who has thriftily made stock from a roast chicken carcass knows, home-made stock is a flavourful base for soups or risottos. The importance of stock in the restaurant kitchen is manifest in the fact that cookbooks by chefs invariably contain recipes for it. Despite its fundamental simplicity, care and attention play their part. If clarity of a stock made from bones is sought, then very gentle simmering over a low heat as opposed to impatient boiling is required, otherwise the results will be cloudy. In order to boost flavour, some stock recipes require a preliminary frying or roasting of ingredients – whether chicken wings or chopped vegetables – to brown them. This has the practical effect of creating the Maillard reaction and so imparting an extra, umami-rich savouriness to the final stock.

The time taken to make stock varies hugely, depending on the ingredients used. Chicken stock usually entails at least an hour, and at least 2 hours when a traditional, tough old boiling fowl is being used. Thin, fragile fish bones, in contrast, require a far shorter time, classically simmered for just 15–20 minutes. In Japanese cuisine, *ichiban dashi*, the core stock, is made in the same time frame from dried kelp (*kombu*) and dried bonito flakes (*katsuobushi*), with great precision required over the water temperature and contact time for each ingredient. Recipes vary in details, but the overall order of the recipe remains the same. First the kelp is used to flavour the water, through either steeping or simmering. The seaweed is then removed, dried bonito flakes are added and the water is brought to full boil. Contact times at this stage are critical. 'If bonito flakes boil more than a few seconds, the stock becomes too strong, a bit bitter, and is not suitable for use in clear soups,' warns the Japanese food writer Shizuo Tsuji. *Dashi* acquires flavour quickly because, as Japanese chef Yoshihiro Murata has observed, time has been put in at an earlier stage: drying both the kelp and the bonito intensifies their flavour.

In contrast, the basic broth used for ramen is slow-cooked, made from meaty pork bones and chicken, boiled for several hours or longer. As the cult 1985 film *Tampopo* showed, getting the broth correct is vital to successful ramen. Rather than the clarity often sought in stock, properly made ramen broth is opaque, pale in colour, possessing a specific fatty texture much valued by aficionados. The heat and the time taken in cooking it in effect break down the bones, creating not only gelatine but emulsifying extracted solids, such as fat and marrow, into the broth to create its cloudy appearance and distinctive mouth-feel. In Western cuisine, a similar approach to breaking down the bones manifests itself in in-vogue 'bone broth', requiring a long simmering of 12–24 hours.

In Chinese cuisine, what is known as a red braise master stock is an important ingredient: an aromatic stock, traditionally flavoured with garlic, ginger, star anise, cassia, rock sugar, light soy sauce and rice wine, used as a poaching medium for pork and poultry. With commendable good sense, after poaching, rather than disposing of the now additionally flavourful stock, it is strained and kept for future use. Each time it is used as a cooking medium, the broth becomes even tastier. In terms of hygiene, to prevent bacteria forming, if kept in the fridge, the stock must be heated to boiling point every few days, then cooled and chilled or, alternatively, safely stored in the freezer between uses. In this way, a master stock can be used for many years. Australian chef Neil Perry of Rockpool, Sydney, uses one that is several years old, and in Chinese restaurants there are reputedly master stocks that have been passed down from generation to generation. 'This is one of the most-used stocks in my kitchen,' writes restaurateur and chef Kylie Kwong in her cookbook *Simple Chinese Cooking Class* of her master stock.

'It is so versatile, as it freezes well and ages gracefully (just like wine).'

Much nineteenth-century ingenuity went into working on ways of providing stock in a quick-to-prepare form. Motivated by a desire to offer a cheap, nutritious food, the German chemist Justus von Liebig developed a concentrated beef extract in 1847, made by boiling beef, then reducing the liquid into a paste. In 1853 the charismatic French chef Alexis Soyer tried unsuccessfully to patent the splendidly named Soyer's Ozmasome Food, made from concentrated meat roasting juices and designed to be used for the making of soup. With the early twentieth century came the arrival of that now everyday short-cut kitchen staple the stock cube, with Swiss company Maggi producing their bouillon cube in 1908, Oxo creating their famous beef cube in 1910 and Knorr producing their bouillon cube in 1912. Stock cubes are sold around the globe, with companies tweaking their flavour profiles for national tastes and offering cuisine-specific products, such as Knorr's tamarind cubes for Thai cooking. Even the quickly made *dashi* has been speeded up: the traditional method used freshly shaved flakes from a hard block of dried bonito, requiring a special tool called a *katsuobushi kezuri*, resembling a carpenter's plane. Nowadays, pre-shaved, pre-packed bonito flakes are sold. Furthermore, packets of *dashi-no-moto* – flavoured with bonito and available in granulated or liquid form requiring simply the addition of hot water in order to make 'instant' *dashi* – are widely used.

At the same time, proper stock continues to occupy a special place in professional kitchens. In classic French cuisine, the time taken to make stock from scratch is not begrudged. 'Indeed,' observed Auguste Escoffier in his 1907 work, *A Guide to Modern Cookery*, 'stock is everything in cooking, at least in French cooking. Without it nothing can be done. If one's stock is good, what remains of the work is easy; if, on the other hand, it is bad or merely mediocre, it is quite hopeless to expect anything approaching a satisfactory result.' Escoffier's words still hold true for French chefs to this day.

'The first thing you learn at cookery school is the making of stock – fish stock, beef stock, chicken stock – because in France we love sauces and stock is the basis for a good sauce,' French chef Pierre Koffmann tells me when I meet him at his eponymous restaurant Koffmann in the elegant surroundings of the Berkeley Hotel, London. A charismatic, bear-like figure, courteous yet quietly formidable, Koffmann's voice retains a rich French accent despite decades of living in Britain. He has championed French fine dining to great effect in his adopted country, notably with his 3-Michelin-starred restaurant La Tante Claire, influencing generations of chefs who have worked with him.

Veal stock, with its delicate flavour and its silky texture, is a key stock in classic French cuisine. It takes '24 hours to make veal stock', says Koffmann, talking me through the process. 'We use veal bones because they give a lot of gelatine as they are high in cartilage. First you roast the bones until they have a very nice caramelized colour. They shouldn't be too dark otherwise it gives a burnt taste, but nicely brown all over, so it takes time to do that.' The bones are put in water, simmered and skimmed often to remove the layer of fat and foam that forms on the surface. Once the stock has been skimmed 'properly', roast vegetables and aromatic herbs are added. The stock is gently simmered over a low heat. 'Never let it boil too fast, because it will be too cloudy,' he exhorts. When Koffmann first came to England in 1970, he remembers, it was hard to find the veal bones he needed for stock. 'There was something called bobby veal – the male baby calves from the dairy industry. It was the size of a dog,' he pats the air expressively beside him, 'they were very, very young, very gelatinous. At the time we had to use that.'

Stock, then, forms the basis for the sauces which characterize French cuisine. Many of the classic recipes that use it involve a further time-consuming process of reducing it considerably to the point of viscosity, concentrating its flavour. 'We have a little book in France called *Repertoire de la Cuisine*; there are hundreds of sauces in it. A sauce is not just the stock; if you serve only veal stock, it's just veal stock,' Koffmann laughs. 'The trick is to incorporate some red wine – Bordeaux – Madeira, brandy . . .' As he talks, I see him mentally flicking through recipes. 'For example, if you want to make a nice Bordelaise sauce, you chop some shallots, sweat them, put in the red wine and reduce it a little bit, then add some stock. You reduce it until if you take a spoon, for example,' here he reaches forward to take a spoon from the service setting at the table to demonstrate, 'you dip the spoon in the sauce and with your finger make a line through the sauce on the back of the spoon. If the line stays there the sauce is ready. It's got to be the right consistency, the sauce, if it's too watery that's not good. If too sticky, you don't enjoy it,' he declares emphatically. 'You have to taste it a lot. That's very important; every time you make a sauce you taste three or four times. If you reduce it too much, you ruin the sauce. The meat must be good too, mind! The sauce is there to complement it, not to hide it.'

Stock-making, including veal, plays an important part in Koffmann's kitchen to this day, but it is something that he feels is disappearing from contemporary professional kitchens in

Britain. 'It's quite an expensive job to make stock, and then when you make a red wine sauce you have to use a lot of wine and the cooking time as well, so a lot of young chefs, they don't make it any more. They might make Béarnaise, which is quick, but the long-process sauces, they are tending to disappear.'

THE SLOWNESS OF SOURDOUGH

In the world of bread, sourdough has achieved a special status, being seen as the ultimate 'slow bread'. This way of making raised bread is ancient, dating back thousands of years. The method is a popular one in many countries, used to produce Russia's and Germany's rye breads and France's *pain levain*. Sourdough is made from dough that has been fermented by naturally occurring yeasts and *lactobacilli*. To make sourdough, the first step is to create a sourdough starter, sometimes called a 'mother'. This can be made simply by mixing together flour and water and leaving it for a few days in a warm place. The water and the warmth trigger the growth of the yeasts and *lactobacilli*, which feed on sugars from the carbohydrates in the flour, creating lactic and acetic acids that give the starter a sour tang. To make a loaf of sourdough bread, the starter is mixed with dough, set aside to rise, then formed and baked. Traditionally, only a portion of the starter is used in the dough, the remaining portion being refreshed with flour and water and set aside to ferment once more. The cyclical nature of sourdough means that starters can be lovingly nurtured for years; tales of sourdough mothers that have been sustained for decades form part of baking folklore.

Whereas industrially manufactured commercial baker's yeast causes dough to rise quickly, sourdough requires time. This innate slowness makes it flexible and tolerant. As baker Andrew Whitley, a great champion of sourdough in Britain, writes: 'The slower a dough is rising, the longer the "window" for getting it into the oven in good shape. You can bake to your own timetable, not one imposed by an impatient yeast.' While countries such as Germany have a long, cherished sourdough tradition, it is only in recent decades that it has become popular again in Britain. Having fallen out of fashion in Britain in the twentieth century when bakers moved to using commercially produced baker's yeast (*Saccharomyces cerevisiae*), craft bakers have pioneered its revival in recent decades. Nowadays, sourdough is found in supermarkets and bakery chains. The lack of a legal definition of the term, however, means that much 'sourdough' on sale is really 'sourfaux', as Chris Young of the Real Bread Campaign explains. This fake sourdough is made using a little bit of sourdough, or more usually a concentrate or powder, to add tanginess, but also with fast-acting baker's yeast, which acts as the dominant raising agent. 'The stuff they're selling as sourdough is made very quickly, so all the wonderful things that happen during the time that real sourdough takes to come alive just don't have time to happen. There's a company selling a packet mix that, in their own literature, says it will allow you to bake sourdough in just 60 minutes. They also say you don't need to employ skilled staff to make it.' The long, slow fermentation in true sourdough brings benefits other than flavour and texture, reducing the glycaemic index and also, according to reports, increasing digestibility. 'People are being misled,' Young says simply, hence calls by the Real Bread Campaign for an Honest Crust act, which would include a legal definition of sourdough, among other demands.

Housed in a sixteenth-century weaver's house in the village

of Biddenden, Kent, chef-patron Graham Garrett's small restaurant, the West House, has a loyal following, myself included. Despite its Michelin star, the restaurant has an informal atmosphere and there is a genuine friendliness to the service here, carried out by Graham's partner, Jackie, and their son, Jake. Each meal starts hospitably with diners presented with two types of bread – sourdough and hazelnut and raisin – with whipped pork dripping and home-churned butter and marinated olives to nibble on. It looks simple, but the care taken to produce this plateful is considerable. Garrett serves his own home-made bread; each time I've eaten here, I've been struck by how good it is. Down-to-earth, plain-spoken and humorous, Garrett's inspiration came from a meal at David Kinch's Manresa restaurant, where the bread had been the best he had ever eaten. Determined to make his own bread that he could serve with pride, Garrett brought a characteristic obsessive enthusiasm to his quest. He invested much time in working on his sourdough recipe, reading, asking for advice and experimenting: 'It took me about three and a half years to achieve a bread that I like,' he says simply. A clear demonstration of sourdough's capacity to enthral.

As I watch Garrett at work, I realize that interacting with the sourdough in its various forms gives a pattern to his working day. His first step is to refresh the sourdough starter, with the mixture of flours, the ratio of flour to water and the frequent feeding all contributing to create the specific flavour he seeks in his bread. He then begins to make the dough, mixing together starter and water, adding flour, using his hands as paddles to form a soft, sticky mixture that he first sets aside for 40 minutes. The following stage is to turn and work the dough by hand, gently stretching and folding it, which he does at 20-minute intervals, setting a timer as a prompt. With his bread served every day in the restaurant, Garrett is looking for consistency in how it is made, so has developed a reliable method which can also be followed by his staff. Each time he returns to the dough, it has changed slightly: air bubbles form, the texture of the dough alters. 'Look at the shine on it,' he says enthusiastically as he works the dough carefully for the second time. 'It's getting stretchy; it's all about capturing the air.' At the third working, he observes that the bubbles are bigger, estimating that the dough is probably a quarter of the way through. The dough is worked this frequently 'until it's ready', a process that might total 3 or 6 hours. Garrett has a very specific sense of what he is looking for: 'There's a point where it's usable, a point where it's perfect, and there's a point where it's past being perfect. Rather than leaving it until it's really bubbly, I find it better to catch it on the way up.' Once judged ready, the dough is divided, shaped into loaves, placed into baskets, then matured for 4 to 5 days in a large chill room. This maturing period is important for giving the bread a better depth of flavour.

Preparing the bread ahead in this way not only gives it enough time to develop as he wants, but also allows Garrett to ensure a steady supply. Having seen the detailed way in which he works, I am unsurprised to learn that the baking process itself is similarly precise. Garrett places one of the matured loaves in his preheated combi oven, cooking it for around 50 minutes overall, adjusting the level of steam and the temperature at intervals as it cooks. When the loaf comes out, it's set aside to cool before we try it: 'I don't usually like waiting to try things,' Garrett grins, 'but it's better cold than warm, believe me.' The loaf, ridged with deep scores to release steam as it is baked, is a rich brown colour, with a crust that crackles as it is sliced. This is proper bread: stretchy-textured, but not heavy, pocked with holes that express the sourdough's activity, having acidity, yet without being overly tangy, the flavour profile that Garrett has so carefully worked on creating. Even now, thinking back to my time at the West House, the sourdough stands out: the way it was both complex and simple, a satisfying food. What stays with me the most, though, is the expression on Garrett's face as, in his busy working day, he paused and tenderly handled the dough, feeling it, knowing it through his fingertips: how he looked so utterly content.

SOUS-VIDE SUSPENSION

It looks uninspiring: a square metal box. The sous-vide, however, is an innovative, sophisticated piece of cooking kit, offering the ability to manipulate culinary time in ways that were previously impossible. Its avowed admirers include many of the world's great chefs, among them Ferran Adrià of the legendary El Bulli, Heston Blumenthal of the Fat Duck, Thomas Keller of the French Laundry, and French chef Joel Robuchon of L'Atelier de Joel Robuchon. Like the traditional bain-marie, the sous-vide uses water as a medium to transmit heat. Where it differs, however, is that the ingredient being cooked is shrink-wrapped in food-grade plastic (shrink-wrapping preventing any air bubbles forming that would create uneven cooking) and immersed directly in the water bath. Vitally, too, sous-vide technology offers its users precise control of temperature.

As a method of cooking, this temperature-controlled take on a water bath has a sophisticated pedigree: working at France's renowned Maison Trosgrois in 1974, French chef George Pralus developed the use of the sous-vide technique in a restaurant kitchen as a way of cooking foie gras so as to minimize wastage of this costly luxury. Whereas previously foie gras lost 30–40% of its weight when cooked conventionally, he discovered that wrapping it in several layers of plastic and cooking it gently in a water bath reduced weight loss considerably. Dr Bruno Goussault of Cuisine Solutions, an American premium foods company, did pioneering work to develop the sous-vide cooking method and disseminate knowledge of this new technique among leading chefs. Part of the fascination of cooking with sous-vide is the way in which it allows ingredients to be cooked at low temperatures for times which are far longer than normally possible with traditional methods. The quest for 'perfect' eggs has resulted in numerous sous-vide recipes. The ability to control temperature produces intriguing results. Eggs cooked for 35–45 minutes at a constant 147°F have a soft, custardy texture, despite the extensive cooking time. Beef short ribs cooked for 72 hours at 144°F are another classic, producing meltingly soft meat. Sous-vide enthusiasts such as Nathan Myhrvold, principal author of the monumental *Modernist Cuisine*, point out that it is a unique way of cooking, producing textures and flavours that are unachievable in other ways. The fact that the ingredient is sealed means that, in effect, it is cooked in its own juices, giving succulent results with an intensity of flavour. Meat and fish can be cooked in a sous-vide with great precision and uniformity, creating, for example, a piece of steak that is evenly rare throughout, rather than cooked on the outside and rare in the centre. It allows food to be cooked without overcooking during the process, so removing the risk of frequently encountered faults such as dried-out chicken breast or salmon fillet. Sous-vide cooking, however, doesn't brown food, so a quick additional searing afterwards to create the desired Maillard reaction is often added.

Sous-vide machines have become a commonplace in restaurant kitchens and, as their prices lower, they are slowly finding a domestic market too. The flexibility they impart – allowing food to be slowly cooked overnight or made in advance before being quickly finished off – is appreciated, especially by busy professional kitchens. 'A sous-vide allows you to suspend time,' observes Nicola Lando, founder of online food retail website Sous Chef and a keen domestic cook. But not every chef is a fan, with many deploring the loss of traditional culinary skills that an overreliance on water-bath cooking brings and the lack of direct contact between chef and ingredients as they cook. 'I'm against water baths,' declared French chef Pierre Koffmann firmly at a talk he gave. 'No smell, no touching. These young chefs work just with a clock and a probe.'

THE CULT OF SLOW

Long, slow, gentle cooking produces results we relish. Tender meat that yields delightfully as we bite into it. Stews with a depth of flavour that demands savouring and appreciating. Many classic, much-loved dishes require time to make, involving considerable periods of simmering or roasting. In Bologna's famous *ragù*, minced beef is cooked very gently with tomatoes for a considerable period, until the sauce softens and thickens, gaining an umami-rich intensity of flavour. 'It must cook at the merest simmer for a long, long time,' declares Marcella Hazan, stipulating 3½ hours as the minimum – and 5 hours as far preferable. A classic Provençal *daube* features beef placed in a tightly covered earthenware pot and cooked in an oven for around 6 hours. Pakistan's national dish, *nihari*, made traditionally from mutton shank or beef cuts containing marrow, is simmered for 5–6 hours. Ample amounts of time are a vital 'ingredient' in these recipes in order to get good gastronomic results. There are also traditional dishes cooked slowly for religious reasons. In Jewish cuisine, for example, cholent is a stew made for Shabbat (Sabbath), the day of rest, when no cooking is allowed. Reflecting the Jewish diaspora, there are numerous versions of cholent; one classic variant consists of beef, potatoes, barley and beans, placed together with water in a sealed pot. Cholent was historically simmered overnight for 12 hours or more in a communal baker's oven, then eaten for a midday meal on Shabbat after morning synagogue service. 'Cholent has deep emotional significance,' writes Claudia Roden in *The Book of Jewish Food*. 'The smell exhaled when the lid is lifted is the one that filled the wooden houses in the shtetl.' Similarly, Boston baked beans, made from dried beans flavoured with salt pork and molasses, traces its origins to the early days of the settlers and Puritan Christianity, with cooking prohibited from sundown on Saturday to sundown on Sunday. The beans were cooked slowly in the residual heat of the baker's ovens all day on Saturday, to provide food for that evening and the following day.

Many historic long-cooked dishes require specific cuts of meat, which need considerable amounts of time in order to soften. 'Slow cooking is for any meat that has connective tissue or collagen in it, any part of the animal which works hard so needs to stick together,' explains Scottish chef Neil Rankin, the author of *Low and Slow*, a cookbook championing the slow cooking of meat. 'So, breast of chicken doesn't have to do anything, nor does fillet of pig, which is on its back. Anything on the shoulder or the leg will be good; the more they work, the more cooking we have to give them.' Cooking these cuts quickly doesn't give them enough time to break down, with unappealingly tough results. Prolonged cooking at low temperatures, in contrast, gradually breaks down the bonds between the fibres – softening the meat – and transforms the connective tissues and collagen into gelatine, adding a succulent thickness to sauces.

Cuts that suit slow cooking – ox cheek, shin of beef, lamb shanks – are often cheaper, perceived as lower-status, despite their tasty potential. While readily available in traditional butchers, these cuts can be hard to find in supermarkets. This was brought home to me one day last year when I was having to shop for ingredients in London's West End. Wanting to make a stew, I went to three mini-supermarkets on a quest to buy braising beef. All I could find, however, were cuts such as steaks, fish fillets or chicken breast fillets. These food shops, I realized, were targeting commuters, snatching a meal on their way home. Not only did most of their stock consist of ready-meals and pre-chopped vegetables, they had similarly focused on speed of cooking when it came to the cuts of meat they offered. Stir-frying, flash-frying, griddling – these fast-cooking methods were catered for. Thrifty cuts that could be transformed into nutritious, sustaining meals through cooking with vegetables and stock were not to be found on that stretch of London high street. It was a telling realization of how impatient retailers feel people have become in regard to cooking.

At the same time – in reaction to our frenetic-paced lifestyle – slow-cooked dishes have also acquired a certain kudos. When time is a rare commodity, dishes that require it take on an alluring novelty, becoming almost fetishized . . . Menus at fashionable restaurants, I notice, now often add times to their descriptions of dishes, such as '6-hour braised beef short ribs'. In a previous age, the use of time in specific slow-cooked dishes would simply have been taken for granted. Time is now so precious that its use in cooking confers distinction and demands to be brought to our attention. Pulled pork, a dish requiring long, slow cooking, has become ubiquitous, popping up in fast food chains as well as street food stalls and restaurants. Recipes by chefs, such as Jamie Oliver's overnight roasted pork shoulder, which requires 10–12 hours' cooking, have given slow cooking an image boost. What is striking about these recipes, in fact, is that while they take time, they require very little trouble. In many cases, one simply leaves them in the oven to cook. Neil Rankin feels that a dish that requires 12 hours' cooking, so can be started the night before, is simpler to make than one that requires 2–3 hours. Rankin's new London restaurant Temper sees him practicing the long, slow cooking he enjoys. One of its distinctive aspects is the cooking of entire animals. How long does it take to cook a whole pig, I ask him, curious to know. 'About 16 hours in a smoker,' comes his laconic reply.

SMOKE GETS IN YOUR EYES

The appeal of barbecued food taps into something primal – our fascination with fire. In his book *Catching Fire*, Richard Wrangham, a professor of biological anthropology at Harvard, makes the striking case that it was our ability to cook food using fire that helped us develop from habilines (the 'missing link' between apes and humans) into *Homo erectus* around 1.9 million years ago. Cooking, writes Wrangham, makes our food safer, more appealing and easier to eat. Most importantly, he observes, 'cooking increases the amount of energy our bodies obtain from food.' In Wrangham's thesis, the extra energy from cooked food gave the first primitive cooks biological advantages. Our bodies, accordingly, adapted to a cooked, rather than raw, diet. Man-the-Hunter is important, but in Wrangham's outline of human evolution, so, too, is Man-the-Cook. The ability to use fire to create light and warmth played an important role in human development. The idea that our capacity to cook food over fire was a major, shaping factor in our evolution is an intriguing one.

In today's world the ability to create heat – using gas, electricity or microwave energy – at the flick of a switch or the press of a button is taken for granted. Cooking over live flames or hot, glowing coals, however, brings an element of excitement, charged with an ever-present frisson of danger. There is a particular satisfaction to observing a fire 'take', watching a small, flickering flame spread and expand to become several larger flames, with an increase in heat and light. Live fires, whether in a hearth at home or outdoors, have the capacity to draw people towards them, mesmerising them into watching the dancing movements of the flames. The food that we cook over fire has a special appeal. We appreciate the appetising aromas created as it cooks and the unique smoky flavour the process imparts to ingredients. In an age of stoves and microwaves, it is the barbecue that offers us the chance to experience this fundamental way of cooking food.

The practice of barbecuing food – whether using direct heat or indirect heat – is found in myriad forms around the world: bulgogi in Korea, shashlik in the Middle East, yakitori in Japan, to name but a few. Among my vivid memories of living in Singapore as a young child is watching the satay man busy fanning his white-glowing charcoal coals as he cooked skewers of marinated, spiced beef or chicken above it. The hungry anticipation was rewarded with the pleasure of eating freshly-grilled satay, served with peanut sauce, pieces of onion and cucumber and cubes of *nasi impit* (compressed rice). Barbecuing in Western culture tends to have a macho image: the traditional territory of pyromaniac dads cooking in the backyard, accompanied by flames, smoke, high heat and sizzling meat rapidly cooked. Chef and Meatopia UK co-founder Richard Turner is adamant, however, that a gentle, slow approach is key to the craft of barbecuing well.

Grilling on a barbecue is a method associated with tender cuts, often chopped into small pieces as with yakitori, shashlik or satay. Although these cuts will cook quickly, Turner recommends a leisurely approach to the whole business. 'The biggest mistake anyone can ever make with a barbecue is to put the fuel on, light it and cook on it. You should never do that,' he says emphatically. Instead, the fire needs to cook down for at least 45 minutes, so that one is cooking over burning embers. When it comes to the fuel to use on a barbecue, Turner advises buying properly made charcoal, produced traditionally and slowly, which burns evenly and well. This is in place of 'accelerated' charcoal, made by dousing low-grade wood with chemicals so that it burns fast, with the resulting rapidly made charcoal releasing these chemicals when it is lit. As food cooks on a grill barbecue, it should be moved and turned to ensure that it cooks evenly and browns well, acquiring those sought-after Maillard reaction flavours.

It's a Saturday morning in early September in London.

Outside Tobacco Docks, a converted warehouse in London's East End, there's a long queue of people. As I approach, I catch the whiff of smoke and upon entering find the air is thick with it, floating ash and a savoury, hunger-making smell. I am reminded irresistibly of Obelix, Asterix's companion, walking, nose raised, a blissful expression on his face, following the scent of boar roasting over a fire. Everywhere I look, there are stalls, from restaurants including Ekstedt, Hawksmoor, Nopi, all offering barbecue-based dishes. Chefs sharpen knives with nonchalant skill, before carving freshly cooked meat, dividing it deftly into takeaway portions for eager customers. This is Meatopia, a barbecue festival founded in the US, now in its fourth year in the UK and welcoming thousands of enthusiasts. As I walk around, I overhear snatches of conversation: 'Now, that's why you need a brazier', 'I did a really good chicken', 'reminds me of Argentina'. Beards and tattoos are much in the evidence among both the chefs cooking and the attendees; barbecuing it would seem, has become distinctly hip.

Listening to Turner discuss the intricacies of barbecuing and the skill and attention required to do it well – sourcing the right cuts of meat, the fuel, type of barbecue equipment, the cooking process – I see for myself barbecue's ability to fascinate chefs and cooks. Around the world, leading chefs such as Victor Arguinzoniz of Spain or Magnus Nilsson and Niklaus Ekstedt in Sweden are exploring the potential of cooking over fire. The sheer range of dishes offered at Meatopia, drawing inspiration from the Middle East, Korea, Finland, Sweden and North America, shows the creative scope within this type of cooking. Among the various tasty dishes I sampled one in particular stood out: a dish from Francis

Mallman, an influential Argentine chef, noted for his mastery of the grill. Looking at my take-away portion of beef, butternut squash and sliver of onion, topped with a simple chimichurri sauce, I wondered how on earth it would be possible to cut the meat with my little disposable fork. My misgivings were misplaced. The pink-hued beef was astonishingly tender, with melting fat. Every mouthful was savoured. I was struck by the delicacy of its flavour – with the taste of the beef predominant, and just a hint of smoke. This was Mallmann's 'hung ribs' – cuts of meat suspended on a metal framework high above smoking embers, with whole onions dotted among them. It had been cooking gently this way for eight hours.

Americans have a special affection for barbecue. 'Americans love barbecue. They love to eat it, argue about it, and even read about it,' observes Robert F. Moss wryly in his book *Barbecue: the History of an American Institution*. In this part of the world, the word 'barbecue' means something different from the grilling over coals the word implies in other countries. The authoritative Harold McGee defines it thus: 'Barbecuing is the low temperature, slow heating of meat in a closed chamber by means of hot air from smouldering wood coals.' It is *this* version of barbecue that is regarded as quintessentially American.

Barbecued food in the United States, as Moss explains, was initially community food, enjoyed by a crowd of people – usually for free – at public events such as Fourth of July celebrations, political rallies, and church socials. Given how vast the United States is, it's not surprising that barbecue is a very regional affair. While pork is in many places the primary barbecue meat, in Texas – cattle country – it is beef which predominates in the pits. Regional points of difference include firewood – oak, hickory, mesquite – type of meat, cut of meat, how it's prepared, cooking time, how it's served, what it's served with and the type of barbecue sauce. It was during the twentieth century that barbecued food began to be sold rather than given away, with itinerant barbecue cooks eventually moving into what became permanent barbecue restaurants. The rise of fast food chains in the twentieth century, however, saw the decline of traditional barbecue. Despite attempts to create fast-food barbecue restaurants, observes Moss, 'their slow-cooking methods and diverse regional variations were ill suited to the standardized demands of the fast-food industry.' It is the innate complexity of producing decent barbecue – the time, care and attention that goes into it – together with its regional richness which makes American barbecue so fascinating. Recent years have seen it enjoy a huge resurgence in popularity, manifested by the spread of barbecue restaurants outside their traditional regions, popular barbecue competitions and a wealth of TV programmes, cookbooks and YouTube videos on the subject.

One enduring barbecue institution is the Louie Mueller Barbecue, founded in Taylor, Texas in 1949. What was once a neighbourhood eatery now attracts visitors from around the world and is held in respect and affection by barbecue aficionados. 'I remember that first morsel at Louie Mueller,' writes renowned pitmaster, Aaron Franklin. 'The place has great counter service – really nice and friendly – and the staff traditionally cut customers a little taste of beef when they get through the line and up to the counter to order. When I got to the front of the line, Bobby Mueller gave me one bit that turned out to be a whole end cut. I don't know why he gave me that gigantic piece. My eyes must have been bulging, and I might have cried a little bit, but it was so, so good. Sooooooo good!'

Today the business is run by Wayne Mueller, grandson of the founder, who kindly takes the time to talk to me. He is a thoughtful, engaged conversationalist, speaking slowly and deliberately in his deep voice and often pausing to think and muster his thoughts before he answers a question. 'Central Texas barbecue is also called "meat market" barbecue because of its origins,' he tells me. During the nineteenth century, Texas saw the mass arrival of Central European immigrants, bringing with them butchering techniques, sausage-making and a 'nothing gets wasted' mind-set. In a pre-refrigeration era, fresh meat was used up in a number of ways: ground to make sausages, dried for jerky and cooked for instant consumption. 'For large, fatty cuts, they discovered that a low temperature, long duration cook would allow for a low tendering, which tenderised the meat and made it more enjoyable.' The legacy of these origins lives on not only in the cooking methods but also in the order experience, with customers able to order by the slice or by weight, as they did in meat markets, and the staff at Louie Mueller's wrapping their order in white butcher paper.

Among an array of cuts, beef brisket is an iconic Central Texas barbecue meat. This cut, as Wayne points out, is composed of two muscles: the upper pectoral muscle (the 'point'), which is very fatty, and the lower pectoral muscle (the 'flat'), which is lean. 'For half of the brisket, these two muscles are stacked one on top of the other; the remaining portion is just the flat. The trouble is that this creates a mismatch. How do you cook this cut so that both sections have reached their desired levels of rendering, without drying them out in the process?' The solution, as Wayne explains, was the development in Texas of a horizontal rather than vertical cooking chamber, allowing the heat to flow along the meat 'like a river', with the thicker, stacked end placed closer to the heat source. 'Fat is a functional necessity of this slow cooking; say you took a tenderloin and used the same process, you would obliterate it.' Correctly rendering down the fat within the meat is important, with that internal 'basting' key to creating the succulent results Wayne is seeking. Brisket cooked vertically, like a pork butt, maintains Wayne, results in overcooked flat, undercooked point – because the intramuscular fat doesn't render down – that has to be carved out and the meat chopped up.

In contrast, the Central Texas method results in tender brisket which can be eaten by the slice. 'I liken us to culinary alchemists,' says Wayne. 'We take this piece of lead – this tough cut of meat – and through our cooking process we turn it unto black gold.' It is the cooking process itself which tenderises the meat. There is no preliminary marinating to help soften the cut, which is seasoned simply with just salt and pepper. The smoke in which the meat is cooked creates its distinctive flavour. At Louie Mueller's the meat is cooked over post oak, which, as its name implies, grows straight and so can be split easily: 'We use post oak because it's a subtle hardwood – that's not bitter or overly harsh – and mostly because there's tons of it around here!' Over the years, people have attempted to speed up the cooking of barbecued brisket, with one method being to first boil the meat, then add liquid smoke. 'The product at the end of it doesn't remotely resemble what we do,' says Wayne simply.

This method of cooking meat requires a huge amount of time. For Saturday's service, the staff at Louie Mueller start cooking on Friday afternoon. Each brisket requires trimming to even out the fat layer, with the cooking process itself taking around an hour per pound. Between the trimming and the loss of fat and water during the process, each brisket weighs about 50% less by the time it's finished cooking. Perhaps unsurprisingly, therefore, when asked what makes a good barbecue chef, Wayne tell me that it's patience. In the world of American barbecue, Wayne Mueller is noted for the way in which he shares his knowledge, with a number of his protégés going on to open successful barbecue restaurants in other parts of America and, indeed, other countries. Each bit of meat requires cooking in its own right, he tells me. 'I teach how to cook barbecue by touch and feel, which is how I was taught to cook. Our fingers become our gauges. What my father taught me was don't listen to what the temperature gauge is telling you. There are places on a brisket, or another cut of meat, that you need to touch and feel. The springiness, the tension that you feel is a scale all of its own. When it reaches a certain level, then you can be certain that this piece of meat is cooked all the way through and rendered properly.' In a world where, 'instant gratification is the rule not the exception', he is amazed that so many people are interested in learning about barbecue, which requires so much time to be dedicated to the process. It's ironic, muses Wayne, that barbecue is classified as fast casual due to the speed of the service. 'When people come in, our transaction time is about a minute per guest. And then I think what you ate today, what you ordered today took a grand total of twenty-three hours to cook. The irony is never lost on me.'

As I talk to Wayne, his own personal investment in maintaining his family's tradition becomes clear. He was eight years old when his father took over the business from his grandfather and immediately began working there, clearing tables, sweeping floors, washing dishes and taking out the rubbish. He began helping to make sausages when he was nine years old. 'I didn't have a childhood. I didn't get to play with the kids; I worked after school and at weekends. My playtime was swallowed by my father's need for me to help him.' His father, however – fully aware of the demanding nature of running a barbecue restaurant – didn't want Wayne to stay and take over the business, encouraging him instead to leave home and go to college, which he did. 'I didn't know how our family barbecue compared to the rest of the world until I left for college. It was only then I began to realise what Dad does is better than what I've had here or here or here or here.' Wayne worked in the worlds of advertising and sport and, he admits frankly, had no desire to return and take over Louie Mueller.

In 2006, however, the James Beard Foundation recognised Louie Mueller Barbecue as an American Classic, which changed everything for Bobby Mueller, Wayne's father. 'My father was just so humble, He wouldn't even call himself a pitmaster; he was just a cook and that's truly the way he saw himself. With the recognition from the James Beard Foundation, he became very adamant that he didn't want to sell it, didn't want to close it. He saw in his own eyes he had reached a level of legitimacy, that he felt deserved longevity – and so did I. I went back.' Eighteen months after returning, Wayne's father died suddenly and unexpectedly of a heart attack, nine months before he had planned to retire. Wayne suddenly found himself running this iconic business on his own.

It was a hard time and he struggled. Two months after his father's death he had an epiphany about Louis Mueller's and his role in it. 'I was sitting down and staring at this wall and the board where people put up their business cards. I was looking at it, thinking about my time spent here, wondering how many hours have I spent in this place? Countless hours and my father – countless hours. And, then I thought about all the people, every guest who has been here and the wall became a timeline – a spectrum of time and experience. What I came away with is that my piece of this timeline is a razor-thin slice, an insignificant portion of it that's not even measurable. It's not my place at all. I found that this was actually an institution for Texas culture, not just Texas food.' This revelation changed Wayne's approach, allowing him to shed his resentment at the way his life had worked out. Today he sounds contented and philosophic, telling me of the deep gratification he gets from watching a customer who's never eaten Louie Mueller barbecue before eat a piece of their barbecued meat. The sense that he is the steward of a cherished part of America's culinary heritage is clearly important to him. 'The greatest compliment I get is when somebody who's eighty years old, who's been a patron of ours literally as long as we've been in business, they can walk in and go, "Yup, tastes same as it always has and the place looks exactly the same." There's comfort in that.'

DAYS

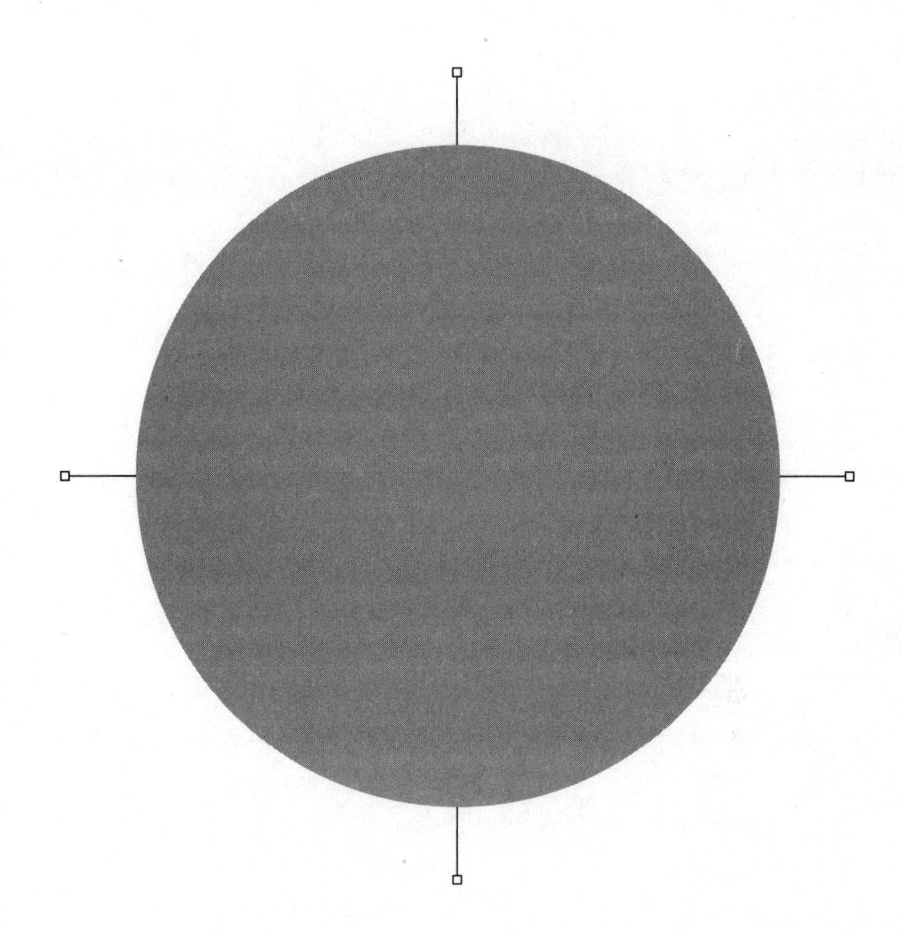

DAYS

DAYS

With days, we enter the realm of preserving food – fighting the decaying effects of time. Historically this was done through transformative techniques such as smoking, curing, pickling and fermentation. Nowadays, of course, technology plays its part in extending shelf-life. The fridge's cool environment allows us to safely keep perishable ingredients – milk, meat, fish, fresh vegetables – for far longer than we could in the past. Traditional preserving techniques continue to be used, valued for the flavours and textures they create and relished in ingredients such as smoked salmon, kimchi and dry-cured bacon.

MELDING AND MELLOWING

Certain aspects of culinary time can be accommodating and flexible, fitted in around busy working lives. Over the decades I've come increasingly to realize that a number of dishes benefit from being made a day before, a fact which allows me to plan ahead. The difference this can make was first brought home to me when I was given an excellent recipe for beef *rendang*, a traditional Indonesian dish made from beef cooked very slowly in spiced coconut milk that I now often make for friends. Ranu Dally, a Singaporean friend who lived for many years in Java, kindly shared her recipe with me; she is a formidably good cook and was very clear that the dish should be made a day ahead. Sure enough, when I taste it on the day I make it, I am always disappointed: the copious amount of onions used dominates, there is a harshness from the turmeric and it all seems rather one-dimensional. Having cooled, chilled and reheated it the next day, however, I am cheered. The flavours have mellowed and melded together, creating a harmonious whole, and it is always eaten with relish by my guests.

The idea that certain dishes benefit from being cooked in advance, while often anecdotal, also manifests itself in cooking instructions. In his authoritative *French Menu Cookbook*, food writer Richard Olney comments of a *daube* (Provençal beef stew) that, 'If anything, the dish improves with a gentle reheating the next day.' Kenji López-Alt's recipe for beef and barley stew in *The Food Lab* contains this telling instruction: 'Serve, or, for best flavour, cool and refrigerate in a sealed container for up to 5 days before reheating and serving.' A traditional recipe for *bigos*, a Polish dish of mixed meat and sauerkraut, recommends making it a day in advance, then reheating it, noting that it will be 'tastiest and "mature" after the third reheating'. The writer Harold McGee muses that 'complex or strongly flavoured dishes may improve with time and reheating'. Refrigeration, which allows us to safely store cooled cooked dishes, plays a part in creating positive results, trapping flavour compounds in protein and starches as they cool. Naturally fatty meats, such as lamb or pork, benefit more than drier cuts, such as breast of chicken. The consensus is that particular types of dishes improve with time, especially those made from powerful aromatics such as onions, garlic, ginger, spices and herbs. With curries, stews, meat sauces and soups, which each contain a number of elements, making in advance allows the diverse flavours to intermingle, for a pleasingly rounded dish. A useful fact when planning a dinner party or faced with cooking for a crowd.

GETTING PICKLES RIGHT

Possessing a characteristic salty, often tangy, flavour, pickles add relish to our food, frequently used to contrast with rich dishes – dainty cornichons served with pâté, the all-important slice of gherkin on a hamburger, petal-like pieces of pink pickled ginger accompanying sashimi. Polish pickled cucumbers, Iran's pickled aubergines, Japan's yellow pickled daikon, piquant Indian pickled chillies . . . our penchant for pickles means that they are found in cuisines around the world. A preserved food, often made from vegetables or fruits, pickles are produced in a variety of ways in different culinary traditions. Ingredients are often immersed in acid, brine, oil or rice bran as well as other keeping mediums. Vinegar is generally the acid used for pickles, its acetic acid content, aided by the addition of salt, acting to prevent the growth of micro-organisms that would spoil the food being pickled. As with many of our preserved foods, our pickling tradition arises out of a practical imperative – a way of safely storing ingredients for future use when they are plentiful. The fashion for using foraged foods among chefs has also seen a rise in pickling, allowing briefly available, seasonal ingredients to be kept in an appetizing form.

Unlike British-style chutneys, which require slow, careful cooking, recipes for pickling can be as straightforward and simple as immersing an ingredient in vinegar, as with Lebanon's pickled turnips, coloured a striking bright pink by the addition of a small amount of beetroot, or Danish chef René Redzepi's recipe for pickled rose flowers. There are, however, more intricate recipes in which time plays an important part in both the preparation and the making. Brining – using salt to draw out excess moisture which would otherwise cause decay, is common in pickling. Blanching or frying, too, for reasons of both food safety and texture, also feature in pickle recipes. Indian pickles often use oils, such as pungent mustard oil, rather than vinegar. Recipes for Indian mango pickle usually follow a pattern: drying chopped mangoes – traditionally in the hot sun – salting them, then storing them in spiced oil. The time taken at each stage, however, varies enormously – from hours to days – with each recipe. The use of oil – olive oil – is central to Italy's *sott'olio* preserving tradition, a way of keeping food which manifests itself splendidly in plates of antipasti. 'The most classic *sott'olio* is eggplant,' says food writer Domenica Marchetti, author of *Preserving Italy*. 'You find it in all parts of southern Italy. Making it involves salting the eggplant to draw out the moisture, letting it dry, quickly pickling it in a vinegar brine, letting it dry some more until it takes on almost a leathery, chewy texture, then putting up in oil.' The salting and pickling are not just to add flavour and texture, but also allow the *sott'olio* creations to be kept for a number of weeks.

Certain pickles require keeping for several weeks. Others, such as *acar*, a traditional, spicy Malaysian and Indonesian mixed vegetable pickle, are ready to eat within a few hours. 'Quick pickles' are at their best eaten just after being made, soon losing texture and flavour. One of the pleasures of making your own pickles is that it allows you to experiment with timings and eat them at the stage when *you* think they are ready. Making a batch of pickled cucumbers, for example, you might prefer to eat them sooner rather than later, when their flesh is less sour; alternatively, you might want to wait for longer for full sourness to be created. The temperature at which pickles are stored will also affect the pickling process, the coolness of refrigeration slowing it down. Every year, Marchetti makes her own *giardinera* pickle in late summer or early fall, inspired by fond memories of the one her Abruzzesi mother used to make. 'I've always liked it from when I was little. My mom's was really good; she just cooked the vegetables so they stayed crunchy in the jar.' Disappointed by the overly aggressive flavours and soggy vegetables found in shop-bought versions, she began making her own, cooking the vegetables very briefly in order to retain their texture and leaving them to cure for at least a week. 'I have to tell you,' she confides, 'that my favourite time of year to eat *giardinera* is in the dead of winter. I like to chill a jar of it and have these crunchy, cold summer pickles, that I made months ago, with a winter dish like a nice roast beef or roast chicken – I love that combination.'

*

An awareness of seasonality is at the heart of Elspeth Biltoft's approach to preserving. Brought up in rural Swaledale in Yorkshire, her childhood was filled with country walks foraging for wild food with her father, the gathered bounty turned into jams, chutneys and syrups by her mother. Inspired by her upbringing, Biltoft set up Rosebud Preserves in 1989. Located in the picturesque Yorkshire village of Masham, its busy workshop is a scene of industrious activity. Much chopping, simmering, jarring and bottling is taking place, as jams, marmalades, chutneys and pickles are made to Biltoft's recipes. The meticulousness of her approach becomes clear as she talks to me about the processes in her soft voice, with a thoughtful earnestness. Pickled onions, a recipe she spent much time and energy developing, are made only at certain times of the year. Biltoft sources small pickling onions from Norfolk, England, just after they've been harvested and 'conditioned' (dried) in sheds for a fortnight, using them from September to November, when she feels they are at their best. The peeled, trimmed onions are blanched just for 1 minute, covered in a brine solution and refrigerated for 2 days – 'Any longer and they become tough,' she observes. The rinsed brined onions are then steeped in a batch of cooked, cooled malt vinegar flavoured with sugar and spices for a further day, then drained and jarred with the reserved vinegar liquor and left for 3 weeks for the liquid to permeate the onions. Retaining crunch, a characteristic of British pickles, is important to Biltoft. 'I want my pickled onions to be snappy to the bite, but they mustn't be too acidic either, that's awful,' she says with feeling. 'It's about balance. You get the spices, with the onions flavoured all the way through.' Another classic British pickle that Biltoft makes is piccalilli, created to an intricate, laborious recipe using a mixture of vegetables, brined and briefly cooked, flavoured with spices and garlic. The secret, she tells me, is to add the different vegetables separately at different stages of cooking. The reason for carefully cooking the vegetables in separate batches is to achieve the texture she wants: cooked through, yet not overly soft, 'a fine line'. Once made, the piccalilli is ready to eat within 3–4 weeks.

As well as having clear views on how to make pickles, Biltoft has clear views about how to consume them. 'People still think, oh, it's preserved vegetables, it'll last for ever. They eat a couple of spoonfuls and shove the jar to the back of the fridge, where it hangs around. If you leave it too long, once it's been opened, it won't taste as good. What I would say is, once you've opened it, get on and enjoy it.'

SUGAR POWER

Sweetness is one of the basic tastes our tongues are designed to recognize, so it is not surprising that sugar-rich foods have long been regarded as luxuries. Sugar is used to enhance natural, perishable ingredients, transforming them into treats while also usefully extending the period for which they can be eaten.

Step into the plush, soft-carpeted environs of London's historic grocers Fortnum & Mason, founded in 1707, during the run-up to Christmas and one of the appealing sights on the sweets counter is a colourful display of candied fruits: red cherries, green pears, black figs, orange apricots . . . The fruits retain their natural form, but have acquired a sticky sheen and the delicately iced appearance suggested by the French term *glacé*. Making crystallized fruit in this way has a venerable history. Dioscorides, a Greek physician (*c.* 40–90 AD), wrote of preserving quinces in honey. Candied fruits were traditionally high-status confectionery, served at stately banquets, and featured at a grand reception for Pope Clement VI in Avignon in the mid-fourteenth century. By the fifteenth century, recipes using sugar, rather than honey, to create them were appearing in Europe. The famous 'sugar plums' of Elvas in Portugal are another historic candied sweetmeat, tracing their origins to the 1500s, when they were made from greengages by nuns in convents and bought by aristocrats.

For centuries, sugar was a rare and expensive luxury, but today it is cheap and ubiquitous. Sugar, like salt, has strong keeping power, hence its use in 'preserves' such as jams, jellies and chutneys. The primary way in which sugar works to prevent spoilage is by dehydrating food through osmosis, drawing water from the inside of food to the outside, inserting sugar molecules in its place.

The making of candied fruit is extremely time-consuming, a process spread out over days. The fresh, unblemished fruit is immersed in what is initially a light sugar syrup, heated and cooled. This is repeated several times – the concentration of the sugar syrup increased over the process – until the fruit is thoroughly saturated with sugar, after which it is dried. *Marrons glacés* are another example of the process, used in France and Italy to make seasonally harvested chestnuts into a costly delicacy. At their best, glacé fruit, while intensely sweet, also retain the flavour of the original ingredient being preserved. These labour-intensive sweetmeats remain luxurious to this very day, associated especially with celebrations. Christmas – when candied fruit and peel appear in Christmas cake, panettone, mincemeat and Christmas pudding – is a festive occasion given a special flavour by their continuing use.

THE SMOKING HABIT

Smoking – used in conjunction with salting and drying – is an ancient method of preserving food. Exposing food to wood smoke coats it in chemicals that slow the growth of microbes which would cause the food to deteriorate. The process also imparts distinctive flavour compounds from the smoke to the food, which are highly valued. Nowadays, much of the food that is smoked is done for organoleptic (i.e. sensory) reasons rather than from the necessity to preserve it. We have acquired a taste for smoked foods.

There are two types of smoking processes: hot and cold. The former is far more rapid than the latter, taking place at high temperatures between 131°F and 176°F that, in effect, cook the food by raising its temperature in the hot smoke. This is a process that can easily be done at home, using either a stove-top smoker or a lidded container such as a casserole dish or lidded wok. Cold smoking is a slower and more complex process, requiring initial curing, that takes place at lower temperatures, typically under 86°F. During this time the food remains unheated, although exposed to smoke. In Britain, cold smoking is classically associated with smoked salmon, a historic luxury appreciated for its texture and taste.

For Gloucestershire-born farmer's son Richard Cook, smoking fish offers a way of conserving not just food, but a local way of life which is very important to him. Ever since, as a small boy, he helped his father catch elvers in the local river Severn, it was fish rather than livestock that fascinated him. As a young man in 1989 he set up a smokery business in an old cattle shed at the back of his parents' house. Speaking deliberately and emphatically, in a rich Gloucestershire accent, Cook tells me how he began buying local salmon and learned how to smoke it. 'I cremated the first few kilns,' he laughs. Today, Cook's Severn & Wye is a noted smokery, with its products, including smoked salmon, used by chefs in high-end restaurants across Britain.

'The essence of preserving something is to take it at its best and then start the preserving process,' Cook declares, fixing me with his direct gaze. Cook's love of fish is clear as he talks with relish of new-season sea trout and the best time of year to eat eels. What is also apparent is how many factors need to be taken into account when curing and cold smoking: size, weight and fat content are all important. The fish must be fat but not too fat, as this would prevent them from taking up the salt successfully. The fact that in Britain we traditionally eat long, fine slices of smoked salmon, produced by cutting across the side of the fish, rather than down into it, requires smokers such as Severn & Wye to create a certain firmness of texture. 'What we need with smoked salmon,' Cook explains, 'is a product that we can slice at around one-eighth inch thick. If the fish is fat and not toned, that makes slicing difficult.'

Cook's personal preference when it comes to smoking styles is manifest in the Severn & Wye cure smoked salmon: it is a '*smoked* fish', he says, emphasizing the first word. When I buy a packet and sample it at home, the texture is noticeably drier than supermarket smoked salmon while the flavour is rich and smoky, in a balanced way, with a real depth to it. The smokery offers a range of cures, created by varying factors such as the amount of salt used, the flavourings, curing time, smoking period and temperature. Their Nordic cure, for example, is a gravlax product flavoured with smoke simply by leaving the salt-and-sugar-cured salmon in the smoky environment of the kiln, but without any contact with actual smoke.

At Severn & Wye, smoking is done traditionally. There are, however, numerous 'fake smoke' products in liquid and powder form that enable 'smoked' products to be rapidly created. There are smokehouses in America, Cook tells me with a grin, that don't have any chimneys. Another short-cut is to brine-inject the fish, rather than dry-salting it, giving a wetter product that, because it weighs more, offers its producer a good return. In contrast, Severn & Wye invest a considerable amount of time in their cold-smoked products, maturing them for up to 5 days after smoking. 'Without that maturation time, we get a product we call "green", that's not got any depth of flavour.' Just as it takes time for the salt to migrate through the fish by osmosis during curing, so it takes time for the smoke to come down through the fish, penetrating into the flesh: 'Both processes benefit from time,' he observes. 'From an organoleptic perspective, it's absolutely key that you allow some time.' As with cheese maturing or the hanging of meat, this maturation period involves loss of weight, so has economic implications for the producer.

The Severn & Wye smokery is a noisy, industrious place. Here the staff work hard, processing thousands of orders each day, transforming fresh fish – salmon, mackerel, herring – into smoked. The labour-intensive process of making smoked salmon begins with huge pallet-loads of 44lb polystyrene crates of fresh, gutted fish arriving from Scotland, Norway, the Faroe Islands. The first stage is to wash the slime from the fish and cut off their heads. Having their heads on when they arrive is an important indicator of freshness: bright eyes and red gills are both good signs. I watch fascinated as whole salmon are sent through a splitting machine that divides them, taking out the dorsal fin, with the two fillets falling off either side of the spine. These fillets then travel along conveyor belts, undergoing trimming by workers as well as passing through a machine that removes pin-bones. As they pass along the line, the form of the fillets visibly neatens. Working quickly, people at the end of the line dip the salmon fillets lightly in salt, which gives them a frosted appearance, and lay them on racks. These are set aside in the cool salting room for a few hours to dry out. Looking at the trolleys of salmon fillets at different stages, I can see that the flesh is indeed losing moisture.

At the heart of the smokery are the kilns – the aroma of the smoke they produce scenting the entire workplace. These use oak chips, moistened so as to burn very slowly and without giving off too acrid and harsh a smoke. On my visit, the door of the largest kiln, big enough to hold two trolley-loads of fish, is opened briefly, giving me a glimpse of the huge, billowing clouds of smoke inside. Once smoked, a process which creates a protective golden layer called a pellicle, the smoked salmon is matured for a number of days. It is then processed for orders, which involves trimming and slicing. The cutting room is noisy, filled with the roar of machinery and the whir of blades, wielded by workers, deftly trimming off the lacquered pellicles, revealing the rich orange-pinkness of the smoked flesh underneath. Slicing-machine blades glide smoothly and efficiently along sides of salmon, creating fine, even slices. The length of the smoking process varies with different cures, a fact which is visible on the skins, which range in colour from light to dark gold depending on the cure. Efficiency in processing orders and delivering is an important part of Severn & Wye's service to its customers. An order placed by a chef at 11 p.m. at night will be at the restaurant by 6 a.m. the following day, provided the item is in stock.

The rise of fish farming during recent decades has done away with much of the seasonality previously inherent in the fish trade. Listening to Cook, though, it is clear how important the seasons are to him when it comes to sourcing and smoking wild fish. Ensuring quality and quantity of supply is a logistical headache for Severn & Wye, though I sense that Cook enjoys a challenge. The mackerel season starts in September, and when I visit in August they are busy talking to suppliers of Marine Stewardship Council (MSC) certified mackerel, discussing issues such as size and fat so that they can, as Cook puts it, 'jump in' at the right moment. When Cook started his career as a smoker, the wild salmon season was far longer that it is now, the shortened season reflecting concerns about salmon stocks. Then, it began in February and ran through to the end of August; nowadays it doesn't start until June. In order to be able to offer wild smoked salmon throughout the year, Severn & Wye have had, therefore, to freeze wild salmon. Speaking with characteristic frankness, Cook explains how unhappy he is at what freezing and defrosting does to the fish: 'You break the cell structures and it becomes a little like a sponge, sucks up the salt really quickly.' On quality grounds, therefore, he has decided not to offer smoked wild salmon out of its season. 'It goes back to what I said about taking something at its best and then preserving it. This year is the first year that we're not going to do wild salmon in the closed season,' he tells me firmly. 'Chefs should only serve wild salmon in the open season and we will support them in doing that, because it's conserving our little bit of England here.'

COOL TECHNOLOGY

For thousands of years humans have known that keeping perishable foods cold was a way of extending the period for which they could be kept without deteriorating. Evaporative cooling, in which water is placed in porous unglazed earthenware vessels so that some of it evaporates, cooling the vessel and the remaining water, has been used for millennia. Naturally formed ice was harvested, stored, then used to chill food and drink. Cool, shady places were also utilized. Perishable, valuable butter, for example, was stored in cavities in the sides of wells, dug deep into the ground. Historic dairies, such as those found on stately country house estates in Britain, are noticeably cool places, even on a hot sunny day. They were deliberately positioned and designed to be such, the use of materials such as marble and stone contributing to the atmosphere. Before the advent of fridges, larders were the places in homes where perishable foods were kept. They were often constructed on the northern or eastern side of a house, in order to receive as little sun as possible. The word 'larder' derives from the Old French *lardier*, meaning a place for meats; this, in turn, derived from the Latin *lardum*, for lard and cured pork.

Nowadays, the refrigerator has become an everyday household item. Three-quarters of homes around the world have one, according to a 2015 Euromonitor report. We now take the ability to keep foods such as milk, fresh fish, poultry, meat and leafy vegetables for a number of days for granted, yet refrigeration technology was developed only relatively recently in human history. In the nineteenth century, important research work was taking place in Europe and America in order to create artificial refrigeration through systems of compression and absorption. Most domestic refrigerators use a system of vapour compression to create a cold environment within them. A fluid refrigerant is turned into a gas through a sudden drop in pressure, becoming extremely cold in the process. This vapour is pumped through an enclosed space (i.e. the fridge), cooling the contents in the process by absorbing their heat. The now heated vapour is pumped outside and compressed to turn it back into liquid, releasing the heat. The liquid is then once more evaporated, transformed into cold vapour and pumped through once again, creating a cyclical process. Keeping food and drink at a low temperature slows down the decaying process by inhibiting the bacteria's growth and reproduction. The temperature inside a fridge is kept at or below 41°F for reasons of food safety (it is known that listeria, a bacterium that causes food poisoning, grows twice as fast at a temperature of 46°F as it does at 41°F).

Refrigeration systems were initially very expensive, used commercially by shippers and the meat industry. The twentieth century saw the rise of domestic refrigerators, the US leading the way. Companies such as Kelvinator and General Motors began producing them in the first two decades of the twentieth century. At first, these domestic refrigerators were luxury items. Streamlined versions, hiding the compressor, were developed during the 1930s, and as prices became more affordable, domestic fridge ownership increased hugely.

Recent years have seen this familiar household staple the subject of various technological advances. Work has been done to enhance crisper drawers, regulating humidity and allowing foods such as fruits and vegetables to retain their moisture, so keeping them in good condition for longer. Some fridges now feature filters that absorb ethylene (the gas released by ripening plants, which speeds up the process), so preventing food from over-ripening too quickly. Blast chillers, allowing, say, a bottle of white wine to be rapidly chilled, are appearing. 'Flexible temperature zones' allow foods and drinks to be stored at their ideal temperature, so potentially extending their fridge shelf-life. Ordinary fridges, for example, are too cold and dry to store cheese well, so the ability to control humidity in a certain zone within the fridge presents interesting possibilities for cheese lovers. Internet connectivity has also entered our fridges – the 'smart fridge' is now being commercially produced. One of the much-touted aspects of this new connectivity is the potential ability to be informed, via scanning technology, when foods are past their use-by date. In a popular scenario, this would allow for replacement groceries to be ordered automatically by your fridge, meaning that one would never run out of milk. Samsung's recently released smart fridge – the Family Hub Fridge – comes complete with a king-size touchscreen set into the fridge door. As befits our surveillance age, the Family Hub also contains cameras that show the fridge's contents in 'real time', enabling you to check what's inside it while away from your home via smartphone and replenish accordingly by ordering online and having groceries delivered. Whether these innovations in refrigerator technology will cut down on wasteful food consumption or increase it remains to be seen.

USE-BY DATE

For centuries, people judged the freshness of ingredients using their senses. With food packaging comes the opportunity for food producers to give their consumers information: the name of the food, the provenance, nutritional value, storage and usage advice. We have also in the UK become used to food labels bearing dates. Today, we seem to have lost our confidence in our own ability to gauge whether or not the food is safe by using sight, taste and smell, relying instead on packet information. In Britain, the 'display until' date is actually aimed at the retailer, rather than the customer. 'Best before' is advice linked to quality, the food still being safe to eat after that date, though in the manufacturer's opinion not as good as it would have been before that date. 'Use by', found on perishable foods such as fresh meat or milk, is advice given for food safety reasons, meaning that the food should be consumed by that date. Research has found that confusion regarding these dates, especially between 'best before' and 'use by', is leading to unnecessary food waste in the UK. WRAP, a UK charity which campaigns against food waste, estimates that around one-fifth of all the food that enters our homes ends up in the bin, much of it unnecessarily. The organization is calling for clearer, consistent information on packaging. It also wants 'freeze before' dates, to encourage people to store food for future use.

In America, where around 21% of food at consumer level is not consumed, the situation regarding dates on food packaging is far more complicated. 'There's no legal framework for date references. There's only one product – infant formula – where the dates are regulated,' Dana Gunders, senior scientist at the Natural Resources Defense Council and author of *The Waste-Free Kitchen Handbook*, tells me. 'Many of the states have come up with their own rules and they're very disparate.' She cites the example of Montana, where milk has to carry a sell-by date that is 12 days after pasteurization, whereas in most other states it carries a date between 21 and 28 days after pasteurization. Given the lack of regulation, food manufacturers have come up with their own practices and phrases and the result is confusion. 'Many customers don't understand the dates – and how would they when there is no standard definition? Many are throwing food out on the date because they think it will have safety issues.' The solution is complex, Gunders feels, requiring education of the public enabling them to shop, cook and eat more efficiently, teaching respect for food. 'When you look at countries where food forms a bigger part of people's budget, then consumers waste far less.' Clear national legislation is something her organization would like to see. 'The irony is that we're trying to at least get where the UK is!'

In a world where the global population is increasing, the issue of food waste is a huge and important one. Food journalist Joanna Blythman is clear on the issue of dates on food packaging: 'Until we stop deferring to supermarkets' own advice on what we may eat and when, those bins will keep filling up.'

FASTING AND FEASTING

Just as human beings are fascinated by food and eating, so, too, are we intrigued by the idea of fasting, denying ourselves sustenance. Eating and drinking is at the heart of survival. Forsaking food and drink for a period, therefore, has long been regarded as having a spiritual dimension. Fasting is found in many of the world's great religions: Buddhism, Christianity, Hinduism, Judaism, Islam. While, strictly speaking, the practice of fasting means doing without any food or drink at all, the word has come to include abstaining from certain foods. Within the Christian Church, fast days were days when the consumption of meat was avoided. The word 'carnival' comes from the medieval Latin *carn*, meaning 'flesh', and *levare*, 'to put away', referring to the celebration held before a period of abstinence, when no meat was eaten. In the Christian calendar, Mardi Gras (Fat Tuesday) is the name given to the day before Ash Wednesday and the start of 40 days of Lenten fasting, reflecting the 40 days during which Christ fasted in the desert. In almost all cultures, centuries-old patterns of religious fasting and feasting have shaped eating patterns and customs. Salt cod, which unlike fresh fish could usefully be stored, was an important food for Christian fast days and remains very popular in Catholic countries. Though today Britain and America are far more secular societies than they used to be, one sees the impact of fasting in customs such as the still prevalent idea of 'fish on Friday' and the continued practice of making indulgent pancakes to celebrate Mardi Gras at the start of Lent. And each day we break our overnight fast with the morning meal we call 'breakfast'.

But fasting, too, has a political dimension. The idea of going without food as an act of protest is ancient, appearing in the venerable Sanskrit text the *Ramayana*. During the twentieth century, hunger fasting was used as a political weapon by people including suffragettes in Britain, Irish Republican campaigners and Cuban dissidents. Mahatma Gandhi, the leader of India's Independence movement and a pioneer of *satyagraha*, a form of non-violent civil resistance, used hunger strikes to great effect to exert moral pressure on the British government. The act of denying one's body nourishment for day after day in order to promote a cause or protest against an injustice is a powerful one.

Despite the physical discomfort of going without food, fasting has also long been undertaken for reasons of health, regarded as a useful way to cleanse the body and clear the mind. The novelist Jeanette Winterson, who underwent an 11-day period of voluntary therapeutic fasting at a German clinic, has written of the 'sense of wellbeing and peace of mind' that it gave her.

Recent years have seen the idea of fasting in the Western world now linked to the idea of dieting and losing weight. Published in 2013, Dr Michael Mosley and Mimi Spencer's book, *The Fast Diet,* puts forward the idea of the 5:2 diet, intermittent fasting for two days (though advocating reduced calorie intake rather than doing without food altogether), in order to, as the book's jacket puts it, 'lose weight, stay healthy and live longer'. In the book, Mosley explains what happens to the body when you fast. The body begins a process called autophagy ('self eat'), in which the body breaks down and recycles old cells. There is evidence that fasting for a short period can trigger the creation of new cells and regenerate the immune system. During fasting, the body first consumes glucose, then glycogen, then begins fat-burning, with the liver producing what are called ketone bodies (energy-rich compounds) from fatty acids as an energy source for the brain instead of glucose. Mosley's research led him to conclude that intermittent fasting puts the body into maintenance mode, bringing with it various health benefits. The book has been a runaway best-seller, garnering much media attention for the idea of fasting. In a world where obesity is a major issue, the ancient idea of voluntarily doing without – cutting down on how much food we consume – has a novel energy to it.

In Islam, the month of Ramadan sees people around the world fasting from dawn to dusk for the entire period. The Pakistani food writer Sumayya Usmani shared with me her memories of Ramadan while growing up in Pakistan. In her family, she explains, there was very much a moral aspect to fasting. When people don't eat they become irritable and bad-tempered, she points out. 'What I was taught, therefore, is when you fast, you learn self-control; when you're hungry is when you're tested as a human.' Furthermore, going without meals offers a chance to appreciate how lucky one is and to empathize with those less fortunate in life who go hungry. For Usmani, thinking about why one is fasting is a vital part of the Ramadan experience. Each day during Ramadan the fast is broken at dusk first by eating a date, then by consuming snacks and drinks that give you a lot of energy. In the north of Pakistan, a Hunza apricot soup is served, whereas in the south, where Usmani grew up, it was a drink made from rose syrup with water and basil seeds. 'Ooh, that first sip was the best thing in the world,' she laughs. 'It tasted like heaven.' Halfway through the period, after the fifteenth fast, people begin to prepare for Eid, the religious holiday that marks the end of Ramadan. Usmani's mother would start to make special foods in advance, stocking up her freezer, while people would begin shopping for new clothes to wear on the festival. On Eid, after prayers, families visit their elders throughout the day to pay their respects. It was a time, Usmani remembers happily, of family togetherness, with food the focus of the celebrations. 'You are offered treats when you visit: *sevian*, lentil fritters, cakes, pastries. Then you have an Eid lunch or dinner with your family, again something rich like a biryani or roast lamb. You get fed from morning to night and you can't say no.'

It is striking that the end of religious fasting periods is marked with a feast, the contrast with the previous abstinence making the abundance all the more appreciated and special. Celebrating an occasion with a feast is a part of every human society throughout history. Religious festivals, state occasions, personal anniversaries – all are marked by bringing people together over food. And the food at these memorable events must be extraordinary. Along with luxurious ingredients and culinary skill, time is one of the elements that characterizes the preparation and cooking of a dish for a feast. Dishes at feasts – such as the splendid macaroni timbale served at a dinner given by the Prince of Salina at his country estate, described in Giuseppe di Lampedusa's novel *The Leopard* – have to be special:

> Good manners apart, though, the aspect of those monumental dishes of macaroni was worthy of the quivers of admiration they evoked. The burnished gold of the crusts, the fragrance of the sugar and cinnamon they exuded, were but preludes to the delights released from the interior when the knife broke the crust; first came a spice-laden haze, then chicken livers, hard boiled eggs, sliced ham, chicken and truffles in masses of piping hot, glistening macaroni, to which the meat juice gave an exquisite hue of suede.

Special occasion dishes that take time and trouble to make are found around the world. *Haleem* is a historic festival dish from Pakistan, made from mutton, barley, lentil and wheat germ. To make it successfully requires simmering for hours until the protein in the meat breaks down and the flavours meld together, with rich, soft-textured results. 'It doesn't look particularly attractive,' Usmani acknowledges, 'but it's very nutritious. A lot of our festivals in Pakistan are related to the idea of giving out food for the poor. *Haleem* is a great dish to give out, because it's a one-pot meal.' For a wedding celebration in Pakistan, she would expect to be offered a proper biryani, *nihari* (lamb shank stew) and rice pudding – 'dishes that show you care for your guests by spending hours laboriously preparing for them'.

Feasting has for centuries been an important aspect of royal custom. A visit to Hampton Court Palace to see Europe's largest surviving Renaissance kitchens gives an atmospheric insight into the hours of hard work and preparation required to feed the royal court during Tudor times. The palace was taken from Cardinal Wolsey by Henry VIII, and was in use until the royal family's last visit in 1737. It was under Henry's orders that the palace was enlarged and made far more splendid – the kitchens were considerably expanded and originally covered around a third of the ground floor. Food production and storage here were spread over fifty rooms. 'They were massive kitchens,' Marc Meltonville, food historian for the Historic Royal Palaces, tells me. 'Even though they look large today, they were much bigger. At least nineteen departments working away. The best way of describing it is as a large hotel kitchen.'

The logistics involved in catering at the palace were considerable, as the kitchen was feeding between 400 and 600 people two cooked meals each day. Organization was vital. A team of administrators called the Clerks of the Greencloth was charged with the not inconsiderable task of ordering foodstuffs, monitoring their arrival at the palace and working out the menus with the cooks. To transform those raw ingredients into cooked dishes, different departments each produced specialized items. Pastry was made in the pastry department and meat was boiled in the boiling-houses, with their great boiling-coppers holding around 79 gallons of water. My visit to the kitchen complex at Hampton Court begins in a courtyard: 'You're standing in the turning circle,' Meltonville tells me. This is where deliveries of raw ingredients were made by cart each day, with the food unloaded and sent to the various storerooms. At the heart of the complex of courtyards and corridors, one finds the impressive Great Kitchen, a huge, high-ceilinged room where meat was carefully and skilfully roasted over huge fireplaces. From here, the food was carried to the magnificent Great Hall and served to those members of the court whose rank permitted them to eat in this space. Those of higher status were privileged to eat grander, more luxurious dishes in the splendid surroundings of the Great Watching Chamber next door.

Among the time-consuming labour of the kitchen staff, sugar work was an important, high-status part. The sugar confectioners at Hampton Court were located above the pastry house, in order to have the warm, dry conditions they needed. Sugar was an expensive luxury in Tudor times, used to create what were called 'subtleties': splendid, intricate constructions which were created as shows of wealth and skill. 'My favourite description is of ships made of sugar with golden sails, brought out on silver mercury, as if sailing on a silver sea,' Meltonville says, with relish. 'The Master of Ceremonies gives a signal and they light a fuse and they all shoot the cannonballs off. That's a feast!'

Even 'everyday' dining at the court was designed to impress: fresh meat, roasted over fires, vegetables and fruit, ingeniously grown by gardeners out of season through the use of shelter, heat, and exotic or unusual ingredients, including spices such as ginger, cinnamon and grains of paradise imported from distant countries. Courtly feasts – elaborate, ceremonious affairs, designed as a show of power and wealth – were accordingly extraordinary events. 'There will always be more food at a royal table than you could possibly eat,' remarks Meltonville. 'That's part of the point of royal dining. And that food will go to feed the poor and that is the final act of charity of your monarch.'

Accounts of the 1520 meeting between King Henry VIII and King Francis I of France at the Field of the Cloth of Gold near Calais give a sense of what was involved when feasting and entertaining at such an occasion. A complete royal kitchen – with bread ovens, roasting ranges and a boiling-house – was set up on site. Hospitality was lavish, the two kings keen to outdo each other. A long list of the fish bought for the event from Henry VIII's sea fisher includes both formidable numbers and unusual items: 9,100 plaice, 700 conger eels, 7,836 whiting, three porpoises and one fresh sturgeon. The cost of provisions for the 48-day trip was over £8,000, a tremendous sum at the time. Conspicuous consumption on the part of the royal court was an important part of the economy.

CHARCUTERIE SKILLS

Salting, fermentation, air-drying and smoking are among historic processes used to transform and preserve meat: examples of human resourcefulness when it comes to not wasting food. From China's 'wind-dried' *lap cheong* sausages to South Africa's cured, spiced biltong or Poland's caraway-flavoured *kabanosy*, the range of meat products is as international as it is diverse. In Europe, the pig, with its useful capacity to put on weight fast and its ample fat, is central to this tradition, particularly in France, with its wide range of charcuterie products made from pork. Pigs were historically slaughtered in the autumn, with most of their meat turned into foods that would last through the cold months. The term 'charcuterie', derived from the French *char cuit*, meaning 'cooked flesh', came to mean both cooked or raw pork products and also the shops that sold these. 'Every small town in France, at any rate in the more prosperous districts such as Touraine, Burgundy or the Île de France, is likely to have more than one charcuterie,' wrote Jane Grigson in 1967, noting that they sold primarily, but not solely, pork products, both ready to eat and requiring cooking. Today 'charcuterie' in Britain is often used to mean only cured (preserved) meat products such as salami, although the range of meat products covered by the term historically includes both fresh and preserved. Charcuterie products are usually regarded as delicacies, with their high prices reflecting the skill and time taken to produce them. Step into a delicatessen and many of the items on sale at the counter will be charcuterie: *prosciutto di Parma*, cooked York ham, *jamón serrano*, pâtés, terrines, chorizo sausages . . . Unlike France and Italy, Britain lacked a historic charcuterie tradition, but the last two decades have seen what has been dubbed the 'British charcuterie revolution'. Across the land, new producers have sprung up, making charcuterie products from meats including pork, but also beef, mutton, venison and goat. Products range from the traditional – *boudin noir*, pancetta, pastrami – to the innovative – duck, pork and Szechuan pepper salami or sloe-gin-flavoured pork rillettes. British charcuterie distributors Cannon & Cannon estimate that whereas in 2010 there were only nineteen British charcuterie-makers, there are now over 200. Such is the 'scene' that, in addition to their wholesale, market stalls and online business, in 2017 they also opened a British charcuterie bar, Nape, in London.

On their farm in Wales, Illtud Llyr Dunsford and his partner, Liesel Taylor, of Charcutier Ltd bring a mixture of tradition and innovation to the charcuterie they make, which is why I wanted to talk to them. Dunsford's connection to the rural landscape around him is deep-rooted, coming, as he does, from generations of farmers who have worked the land in the Gwendraeth valley for over 300 years. He grew up making simple charcuterie, such as sausages, black pudding and bacon, with his grandmother and his mother. The annual slaughter of a pig and its transformation into charcuterie was, he remembers fondly, a very social time, with neighbours and family coming round to help with the work. This experience has informed both what Charcutier Ltd make and how they work. Out of respect to the custom that pigs were slaughtered during the months with an 'r' in their name, offal products, such as faggots, are produced only during the colder months. Based on the couple's travels and research, there is, however, a noticeably cosmopolitan aspect to their range, which includes European charcuterie items such as bratwurst, salamis and *lardo.* A pig's carcass for Dunsford offers him opportunities to create both fresh products, such as sausages that can be made that very day, and cured meats, which require considerably more time to produce.

The level of care they bring to the process begins with the pig itself. The indigenous, white, lop-eared Pedigree Welsh is their mainstay. 'We use it,' Dunsford tells me, 'because it's evolved over generations in its natural environment here in Wales.' In order that the meat has the desired fatty acid level he wants to create in his charcuterie, Dunsford ensures that both the farm's own pigs and those bought in are fed a specific soya-free and GM-free diet. A hallmark of the way they work is to use larger animals, at least 198lb in weight, that offer the cuts required. These consist of three types: a 9-month-old pig, young sows about 13–14 months old that have had one litter, and sows, generally over 4 years old. 'The older the animal the better the quality,' he says emphatically. 'It's the flavour, the colour, the way it develops – everything.' Initially, they bought in older animals specifically for their salamis, but having realized how flavourful the meat was, they now use them across their range.

To make quality salami, Dunsford explains, is very labour-intensive, because of the amount of careful trimming of sinew, connective tissue, bone, skin, soft fat and glands required for cuts such as shoulder, in order to use only meat and fat. Their traditionally made salamis are first fermented for around a week, then matured in a room with constant temperature and humidity levels, a process that takes around 3–4 weeks for snack salamis, up to 6 months for a wider diameter. The cheapest industrially produced salamis are made using glucono delta-lactone (E575), an acidifier, and thermal processing, accelerating production to as little as 24 hours.

One of Charcutier's best-selling products is their dry-cured black collar bacon, inspired by the historic Bradenham ham cure, a recipe dating back to 1781. To make it, the pork is salted and flavoured with treacle, sugar and herbs ('The messiest thing we make,' laughs Dunsford), salted for 7 days, then dried for a further week. Their classic dry-cured bacon is salt-cured for 7–10 days, then dried for between 7 and 21 days. As we talk, the importance of time in the processes Dunsford is describing becomes clear. The long drying process for the bacon is vital: 'With the moisture loss, it increases the concentration of flavour, drawing out the porkiness of the meat.' Most industrially produced bacon is made in 24 hours: the pork is injected with brine, then processed in a vacuum tumbler to soften the fibres and push the brine through the whole cut. This soft, floppy-textured bacon 'weeps' liquid as it fries in the pan, stewing in moisture rather than frying, while rashers of properly dry-cured bacon take on an appealing crispness, as their fat turns a deep golden colour.

The product that takes Dunsford and Taylor the longest amount of time to produce is the ham they make for personal consumption, a country ham made as was traditional in Dunsford's family. The pork leg is salted, left for a certain number of days, depending on the size, rinsed of salt, then hung to dry and mature for around 9 months. This type of ham, found in France, Scandinavia and the southern states of America, Dunsford explains, is usually eaten boiled. 'It can be eaten raw, but it doesn't have the same subtle nuances that some of the Spanish or Italian air-dried hams have.' The initial plan for Charcutier Ltd, he tells me, had been to make air-dried hams in the UK, but the logistics and time involved would have been considerable. 'I worked out the other day that had we started as an air-dried ham business with a piglet it would have taken us 7 years before we could sell the first ham!' Fascinated by the science of charcuterie, Dunsford remains committed to pursuing this vision, prepared to put in the careful research, experimentation and time in order to achieve their dream. 'It's easy enough to make an air-dried ham, but to make a very good one is hard. We're still about 5 or 10 years away from doing that,' he says, philosophically.

FERMENTING FERMENTATION

Fermentation plays a crucial role in the creation of many of the foods we enjoy: cheese, bread, chocolate, soy sauce, olives, beer, wine . . . It is what adds bubbles to champagne and gives charac-teristic flavour to an array of foods from salami to vanilla. Fermentation is defined as chemical transformation through the actions of micro-organisms such as yeasts, bacteria or moulds. In foods fermented by lactobacillus bacteria, the bacteria feed on sugars, creating lactic acid which gives the distinctive sour tang found in fermented foods such as yoghurt or kefir. In alcoholic fermentation, yeasts consume sugars and excrete alcohol and carbon dioxide. By its very nature, fermentation requires time.

Human beings have fermented foods and drinks for millennia. Beer-like beverages, created by 'wild fermentation', using naturally present wild yeasts in the air, are thought to have been made since Neolithic times. Fermentation has long served a range of culinary purposes. It is used to preserve perishable foods, but also to create digestibility, flavour and texture and enhance nutritional content. Many historic delicacies are made using fermentation processes that require substantial amounts of time. In Asia, these include Korean kimchi, Chinese *pu-erh* tea and Japanese *katsuobushi*, the dried, fermented skipjack tuna which is used to flavour *dashi* (stock). Sauerkraut, so important in Eastern European cuisines, is made from fermented salted cabbage. A distinctly pungent fermented delicacy is Swedish *surströmming*, produced from fermented herrings. Nowadays, *surströmming* is usually available in canned form, fermented within the can for 6 months to a year, a gaseous process that causes the cans to bulge accordingly. One of the factors that speeds up fermentation is warmth. There is a tradition among tropical countries of using batters that have been fermented for a few hours, taking advantage of the natural climate to create this effect. India's crisp *dosai*, soft *idli* and Ethiopia's distinctive, flannel-textured flatbread *injera* are all examples of this, with their tangy flavour part of their appeal.

Despite our long history of using fermentation to enhance our food, the advent of more recent ways of preserving food, such as canning, refrigeration and freezing, has seen people moving away from it domestically. The charismatic, articulate 'fermentation revivalist' Sandor Katz is on a mission to reintroduce people to the joys of fermenting food and drink at home. Much of his time is spent talking and teaching fermentation techniques, encouraging people to overcome their fear at the idea of bacteria. While fermentation does involve time, many fermented foods and drinks simply require to be started, then left in the right conditions. Furthermore, the time involved depends on what one is making. 'You could make yoghurt in less than 4 hours,' he says. 'If you want to make Cheddar cheese, that will take longer.' Talking to Katz, I feel empowered by his words. He advocates a flexible, practical approach, based on personal preference, when it comes to the use of time in fermentation. In Russia, sauerkraut was traditionally fermented a month to 6 weeks; however, there is no hard-and-fast rule that says one has to ferment it for that period. As a first fermentation project, Katz often recommends making sauerkraut. What he tells people to do is taste it after a few days, pack it down again, then taste it again. 'That way they get familiar with the spectrum of flavours that's possible,' he explains. 'Many, many people will prefer it after a week to after 6 weeks. There's no virtue in waiting 6 weeks if you like it better after a week!' For people with busy lives, Katz points out, spending just 15 minutes cutting up a cabbage and putting it into a jar to get something nutritious and tasty that you will enjoy eating for weeks is an efficient use of time. 'It's just that there's an aspect of waiting. You have to make it a little bit in advance, anticipate your desire.'

My conversation with Sandor Katz galvanizes me to start exploring the world of fermentation for myself. I begin by making water kefir, a probiotic water-based drink created using a microbial culture. The first step is to order a packet of dehydrated water kefir grains from an online website recommended to me by a friend who makes milk kefir. The process, it turns out, is remarkably quick and easy. I mix sugar with water and place this together with the grains in a large container (plastic, rather than glass, is recommended, as being safer in case of undue fermentation activity). As suggested, I add lemon slices and dried fruit for flavour, then simply set the mixture aside. The theory is that as the culture feeds on the sugar, the mixture will begin to ferment.

A day later, I glance at the mixture and spot a tiny bubble rising to the surface; just this one small sign of vitality makes me feel excited. After 48 hours, a transformation has taken place: the mixture has turned pale gold in colour, cloudy and fizzy, with a thin, white, frothy rim, and a perceptible sour smell. Upon straining the kefir and tasting it, I am pleasantly surprised. The flavour is delicately sour-sweet and the fizziness is noticeably subtle, a world away from the forceful bubbles found in Coca-Cola. The drink I have made is tangy and refreshing.

Among the thrifty pleasures of making my own kefir is the fact that, after the initial outlay, the grains can be used over and over again. As the grains become more vigorous, the cycle of creating water kefir speeds up, with the results now ready for me to enjoy the following day. I experiment with flavourings. I discover that adding slices of lemon and root ginger plus a couple of dried unsulphured apricots gives a deep golden colour with an apricot sweetness and citrus notes, and this becomes my favourite combination. Because they are being nourished every day, the grains grow over time. This has an impact on the flavour, making it far more sour more quickly, so I learn to add additional water to create the level of 'tang' that I want. My early-morning routine now consists of first feeding our cat – who paces back and forth impatiently waiting for his plate to be put down – then topping up the bird feeders in the garden and straining and refreshing my water kefir. Each of these is a nourishing act which gives me satisfaction.

For Katz, the rhythm involved in fermenting food and drink is something pleasurable, to be enjoyed. He cites the making of kombucha, using a kombucha mother to begin with, following a 10-day process, so every 10 days you are brewing fresh tea, adding sugar to it, draining out your old kombucha. In this way, your life becomes patterned through making something you will enjoy consuming. With many of my friends baking sourdough bread or fermenting kombucha or kefir, there seems to be a new and real interest in fermenting food. Katz feels that this is part of a bigger picture: 'I think these are all ways that people are trying to satisfy this hunger for a feeling of connection to the foods they're eating.'

ESSENTIAL SALT

Salt (sodium chloride) is essential for human life. Our bodies need it to transmit nerve impulses, to enable our cells to function normally and to maintain fluid balance. In culinary terms, salt is also of great importance. Long ago, humans understood that salting food helped to preserve it, allowing them to store valuable but perishable foods for future consumption. Salt's capacity to dehydrate – drawing out water from food by osmosis and inserting salt molecules into the interior – is central to its preserving properties, as this deprives spoilage bacteria of the moisture they need to survive and thrive. Around the world, one finds delicacies created through the use of salt's keeping properties: Italian anchovies, French Bayonne ham, Chinese salted duck eggs. Furthermore, we humans enjoy salt. Our tongues contain receptors that detect saltiness. It is a taste that we relish and crave.

Salt's importance in creating foods that could be successfully stored and transported is epitomized by the remarkable history of salt cod, a food which due to its keeping qualities has travelled round the globe. Produced by salting and drying fresh cod, salt cod was an essential ingredient in Catholic Europe, eaten on fast days. The discovery of huge shoals of cod fish off the North American coast resulted in salt cod, which could be kept for months, being shipped to regions around the world, including the Caribbean and Europe. Salt cod features in Caribbean dishes such as Jamaica's saltfish and ackee. Portugal is particularly known for its love of *bacalhau* (salt cod), nicknamed *fiel amigo* (faithful friend) and used in numerous recipes there. The drying process transforms moist, soft-fleshed cod into a formidably tough, dry food, resembling planks of wood with a strong, penetrating smell, and requiring sharp knives or machines to be successfully cut. In order to make salt cod edible, it is necessary to both rehydrate it and remove the excess salt. Recipes for salt cod dishes, such as *bacalhau com todos* (salt cod with everything), therefore, stipulate a lengthy soaking time, generally between 12 and 24 hours depending on the size and thickness of the piece. During this soaking, the water should be changed four or five times in order to freshen it. In Spain, where salt cod is also enjoyed, traditional food markets often feature special marble counters. These are inset with shallow basins created especially for soaking *bacalao*, offering customers short of time the chance to buy it pre-soaked and therefore ready to cook. For those unable to source salt cod, there are recipes for home-salted cod. This is a comparatively quick and easy process that takes place in a refrigerator and requires only 48 hours, removing some of the fish's moisture and giving it a salty tang.

Time is also required for brining, a culinary process based on salt's properties in which food is immersed in a brine solution before cooking. Brining is used to prepare dishes such as corned beef, which is made by soaking beef brisket in brine for several hours before cooking it. In recent decades preliminary brining of turkeys for some hours before roasting has become popular for Thanksgiving dinners. The advantage of brining, explains American chef and broadcaster Chris Kimball, is that 'it helps the proteins to hold on to water during roasting, giving more flavour from the salt and more juiciness'. Brining helps transform dry, tough turkey meat into moist meat and so has taken off in kitchens across America. (There are downsides, however. Kimball cautions that it can result in 'wet' meats and also over-salty ones.) Dry-brining or salting – where salt is rubbed on the surface of the ingredient, which is then set aside in the fridge overnight or for up to 48 hours – is also increasingly considered a good option for home cooks seeking to enhance their roast chicken.

For millennia, salt was both useful and precious. A reminder of its commercial value exists to this day in the word 'salary', with its roots in the Latin word *sal*, meaning 'salt'. It has been processed by human beings for thousands of years. There is archaeological evidence of venerable saltworks found in countries including China and Romania. Salt was essential to trade in many countries, creating historic routes such as Rome's Via

Salaria. Natural salt is found in two forms: in underground deposits created by dried-up seas (rock salt) and in sea brine (sea salt). Mining salt from underground was arduous, perilous work. The spectacular Wieliczka Salt Mine in Poland – a UNESCO World Heritage site – offers striking testimony to the sheer hard work that went into mining salt. Dating back to the thirteenth century and actively producing table salt until 2007, this vast, cold, labyrinthine underground mine contains extraordinary chapels carved out over the centuries by the devout miners who worked in this dark, dangerous place. Today it is a major attraction. Over a million people visit the mine each year to admire these sights hidden away beneath the surface, among them the Chapel of St Kinga – a church measuring 177 x 59 feet, 39 feet high – totally carved from rock salt. Nowadays, most rock salt is extracted by a process called solution mining. Water is forced down shafts deep into the ground to reach underground salt deposits, which dissolve in the water. This saline solution is then pumped to the surface and the water is evaporated using a process called vacuum evaporation. Rapid concentration of brine results in the tiny, cubic crystals that we know as granulated salt. Rock salt for human consumption is refined, a process that includes adding anti-caking additives to prevent the salt crystals from sticking together.

Evaporating sea water using the natural power of the sun to create salt is an ancient way of producing it. A visit to the historic Trapani salt pans in Sicily is an evocative experience. In local folklore, the Phoenicians are credited with bringing the harvest of salt to this part of the island. The combination of naturally shallow, calm coastal waters in the lagoon, hot winds and sunshine offers a suitable natural environment for this work. The salt of Trapani is mentioned in a book written in 1154 by the

Arab geographer Al-Idrisi. In the bright, harsh sunlight, shallow ponds of sea water stretching out before one into the distance mirror the cloudless blue skies. The stillness of the scene reflects the slowness of this process of natural evaporation, which follows an annual cycle. Sea water is pumped into the ponds in April, then moved through the pans until it reaches the final crystallizing ponds in which salt crystals form, resulting in a white layer on the surface of the water. From June to September the resulting salt is gathered by hand, piled into mounds – which glitter in the light – to dry out, then crushed, ground and packed. Highly prized *fleur de sel* is a specific product of salt pan evaporation. It consists of a layer of fragile crystals that forms on the surface of the sea water in calm, still conditions. Before this layer sinks, it is carefully raked off the water by hand. The resulting unrefined salt – highly brittle and soluble – retains the natural minerals from the sea water.

Given Britain's uncertain climate, it is not surprising that sea salt production in the UK takes place indoors. In 1998 on the Isle of Anglesey, off the Welsh coast, husband-and-wife team Alison and David Lea-Wilson enterprisingly set up a new sea salt business, Halen Môn. 'We literally just took a saucepan, took it down to the beach, filled it up with sea water, put it on the Aga and made salt. Yes, it was a bit grey, but we knew we could do it!' Alison Lea-Wilson recalls. From that first, basic step, the Lea-Wilsons spent a lot of time and energy working on 'perfecting' the process, finding the right shape and texture. 'The difficulty was in trying to get the crystals to form correctly. We wanted crystals that were soft enough to pinch between your fingers. Crystal growing involves a mixture of humidity, the temperature and the concentration of the brine, and the time you allow it to form.' At Halen Môn the sea water is pumped ashore, filtered through sand and carbon, then boiled under vacuum. The concentrated brine that's produced is then flowed into shallow, open tanks called crystallizers, where heat lamps drive off more water in the form of steam. The brine begins to form salt crystals, which, as they become larger and heavier, fall to the bottom, with the next surface layer crystallizing in its place. The formed salt crystals are taken out and rinsed in brine to get rid of some of the calcium carbonate, which would otherwise give a chalky taste, creating Halen Môn's characteristically shiny salt. The whole process, from sea water to drying and packing, takes 11 days. 'Most of it is done quite slowly,' Lea-Wilson explains. 'It's very labour-intensive; you can't hurry it along.' Their carefully produced salt is highly regarded by cognoscenti in the world of food, with famous chefs such as Heston Blumenthal of the Fat Duck in Bray among their customers. I, too, am an admirer and user. A small, beautiful horn bowl, given to me by a dear friend, containing Halen Môn's distinctive bright white, irregularly shaped salt sits by my stove. I like the idea that salt – a remarkable substance, which was so prized for so many centuries – continues to be made to this day with that level of obsessive attention.

CORN

There is a complexity to corn, one of our staple grains. Within the species *Zea mays* – corn or maize, as it is also called – there is an abundance of varieties. An aspect of this diversity is that corn exists in a range of natural colours, including white, yellow, red, blue and black. When it comes to categorising corn, it is usually divided into five broad groups based on starch content and appearance: flint corn, which, as the name suggests, is high in hard starch; popcorn, another very hard corn; dent corn or field corn, with softer kernels than flint and a distinctive dented appearance to its kernels; sweet corn high in sugar; and floury corn, composed largely of soft starch.

Historically, these different types of corn are used in a range of ways in the kitchen. Flint corn, with its protective outer layer of hard starch, dries and stores well; in order to allow these tough kernels to be consumed by humans, it is usually ground into flour or meal. Widely grown dent corn is similarly milled into grits, meals and flours. This variety of use, though, extends well beyond the kitchen. Unlike the two other staple grains, rice and wheat, much of the corn that is grown is not destined for direct human consumption. Instead, it is used for animal feed – in the United States, the world's largest corn producer, the grain is fed to livestock and poultry, such as cattle held in feedlots, in order to make them speedily gain weight – to make biofuel, to make corn syrup, and, through the extraction of corn starch, in manufacturing.

Once a wild grass, the corn we know today comes originally from Central America. It is thought that the domestication of corn occurred in Mexico around nine thousand years ago, with a wild grass called teosinte posited as its ancestor. Appropriately, therefore, in this part of the world, corn became an essential crop. 'All the peoples of Mesoamerica, from the Maya to the Aztecs, placed corn at the centre of their culture,' writes chef and food historian Maricel Presilla, in her seminal work *Gran Cocina Latina*. As evidence, she cites the corn myths, such as the creation myth of the Quiche Maya of Guatemala, which tell of Paxil, a land of abundance and corn, and also the historic *metate*

– the three-legged, stone grinding slabs on which corn was ground – found by archaeologists throughout the region.

Drying corn in the sun allowed this fresh cereal to be stored and kept safely, making it an extremely useful ingredient. One key aspect of corn's culinary usage in Latin America is the ancient practice of nixtamalization, a process in which dried corn kernels are boiled with ash or lime. The word is Aztec in origin, derived from *nextli*, the Nahuatl word for ashes. Cooking corn in this alkaline solution meant that the tough outer layer on the kernels could then be easily rubbed off. The process also had the invisible, but important advantage of making the corn more digestible and improving the bioavailability of vitamin niacin (Vitamin B3). Communities that adopted corn as a mainstay of their diet, but without practicing nixtamalization, suffered from the disease pellagra (caused by a deficiency of niacin) as a consequence.

To this day, corn remains a fundamental ingredient in the cuisines of Latin America. The range of corn varieties and the inventive culinary ways in which both fresh and dried corn are used are striking. It is grilled on the cob, used in salads, soups and stews, fermented to make drinks and – a fundamental use – dried, nixtamalized, ground and mixed into *masa* (dough) from which tortillas and tamales are made. The work required for dishes is striking. In her recipe for *Pozole*, a Mexican soup or stew made from nixtamalized dried corn, Presilla describes how the Mexican cook 'is willing to cut off the germ end of every single kernel of corn in order to eliminate the slight bitterness of the germ and make the kernels flower when cooked again.'

The preparation of corn in Latin American cuisine requires time and care, Presilla tells me. 'If you were a Mesoamerican woman in a village where basically you lived on tortillas or corn gruel, your day would be determined by the preparation of the dried corn. You are making bread and flour out of dried corn and your preparation begins the night before. You have to take the kernels off the cob, you have to soak the kernels in water with lime or ashes. You can't hurry the process. In the morning, you have to take the skin off the kernels, you have to grind the masa, you have to make your tortilla and the process is repeated through the day.' This patterning of the lives of people who depend upon corn as a staple exists to this day. 'This still happens in many places in rural areas; that's the way the day's chores are organised and it is the way that corn is at the heart of their lives.'

Unsurprisingly, the arrival of the new season's corn is something to be celebrated in Latin America. In Mexico, the harvest of fresh corn is marked with an *elotado*, which sees people gathering together to cook and eat it in the cornfields. Presilla reminisces happily of the annual July carnival held in her Cuban hometown, which saw her community feasting on the new season's fresh corn. 'People would make fresh corn tamales from the starchy corn. You had to take the husks off, scrape the corn, grind the kernels, make a sofrito sauce, wrap the tamales and boil them. It was a time of the year for feasting, for enjoying, for getting the family together for a process that is time-consuming.' She tells me of how the summer before, having discovered a farmer in New Jersey growing a starchy variety of corn used by Cubans, she took her two young goddaughters to pick the corn and then use their harvest to make these fondly remembered, traditional fresh corn tamales. 'It took two full days,' she says laughing. 'One day to get the corn, another day to produce the tamales. It really was time consuming, but it was so special. That's how we used to do it – all these things that require transformation, the essential element is time.'

Cooking times for corn vary according to a number of factors such as the variety of corn being used, whether it is fresh or dried, and, of course, what is made from it. The traditional making of popcorn sees those small, stony, granular kernels heated with a little oil in a lidded pan over direct heat. The heat converts the water inside the popcorn into steam, creating pressure within each kernel. Harold McGee in *McGee on Food & Cooking*, his masterly exploration of the science of cooking processes, explains the process with characteristic clarity: 'The softening of starch and protein continues until the internal pressure approaches seven times the external pressure of the atmosphere, at which point the hull breaks open. The sudden pressure drop within the kernel causes the pockets of steam to expand, and with them expands the soft protein-starch mixture, which puffs up and then stiffens as it cools off, becoming light and crisp.' This dramatic transformation – complete with percussive sound effects from the exploding kernels (listening is an essential part of the successful making of popcorn) – takes just a few minutes and the results are best eaten freshly made.

In Italian cuisine, *polenta* made from ground corn is a traditional and popular dish; the ground corn is cooked in water until softened. In the best Italian way, there are numerous regional variations regarding the ratio of ground corn to water and the resulting thick or thin texture. Historically, polenta was made in a copper pot called a *paiolo*, stirred frequently as it cooked. Patience and polenta have traditionally gone hand in hand. A *paiolo* and the long wooden spoon used to stir the ground corn mixture are clearly depicted in the 1740 painting 'La Polenta' by the Venetian artist Pietro Lunghi, which shows a golden mound of freshly made polenta triumphantly poured onto a serving dish. This slow, attentive way of cooking polenta continues this day. The Italian chef Giorgio Locatelli writes affectionately of how the cooking of polenta was a Sunday ritual for his Lombardian family. 'Even after watching my dad make polenta so many times, it is still considered that I am only serving my apprenticeship. For forty minutes, he is there in his big apron, tending the pot, folding and stirring with the big, cured wooden stirrer that my granddad also made, making sure that the polenta bubbles slowly, like a volcano – erupting every now and then.' Today, of course, as with rice and pasta, there are speeded up versions of polenta available. Instant polenta, made from part-cooked polenta, sprinkled into a pan of boiling water, thickens within just five minutes.

A similar, slow, conscientious cooking also characterised the traditional cooking of grits, the dish made from coarsely ground corn kernels that comes originally from the southern United States. When it comes to the cooking of grits, 'Stir, stir, stir!' is one of chef Sean Brock's rules in his restaurant kitchen. In Husk, his acclaimed Charleston restaurant, Brock champions his rich heritage of Southern food. As a boy, he grew up on instant grits, made in minutes from a packet. It was eating a bowlful of slowly cooked grits, made from freshly milled heritage corn, which came to him as a revelation of what grits could be. The slow-simmered grits on the menu at Husk are made from heirloom corn varieties, such as the Jimmy Red variety, stone ground in small batches, with the focus of both the mills and the restaurant kitchen being on capturing the subtle flavours and textures that corn can offer.

Dent corn or field corn appears regularly on breakfast tables in the form of corn flakes. In her thought-provoking book *Much Depends on Dinner*, Margaret Visser wrote memorably of the appeal of corn flakes: 'They are as crisp and snack-like as potato chips, sweetened and salted like junk-food, yet hallowed by Milk, which always says "Mother". In North American culture nothing bathed in fresh milk can be threatening or bad.' It is, however, another type of corn, sweet corn, which is seen as quintessentially American – from the bucolic ideal of a plump, soft-tasselled ear of corn, ready to be picked and shucked, to the jolly Green Giant proudly surveying a field of corn on the label of a can of sweet corn, it is this that comes to mind when one thinks of corn and America. Sweet corn is eaten immature, harvested at a stage when the kernels have soft wrinkled skins and are rich in stored sugars. It is treated as a vegetable rather than a grain, preserved not through drying but by canning and freezing. Presilla remembers how, when she came to live in the United States from Cuba, she was surprised by the lack of choice in the corn on sale. 'I had to learn how to cook with sweet corn and so learnt how to cook down the masa for the tamales to get rid of all the water. The thing is,' she points out, 'you cannot get rid of the sweetness, so then you have to get used to the idea of sweetness with savoury, which I don't dislike, but it's very different.' When she ate the fresh corn tamales made with starchy corn she had made with her god-daughters, 'It really reminded me of home. There was not a smidgeon of sugar and South American corn is like that, no sweetness.'

Sweet corn on the cob was, classically, a vegetable which epitomised the idea of freshness: 'When I was a boy, they used to say that you should put your pot of water on to boil, before you picked the sweet corn,' an American friend told me. Once picked, the ideal was to cook it within minutes. There was a good reason for this dictum. The stored sugars, which give that characteristic sweetness, used to quickly turn to starch. 'Whatever you do, don't pick (or buy) corn and then leave it sitting on the kitchen counter to wilt and lose its sugars in the heat of the day,' advises Marian Morash firmly in her *Victory Garden Cookbook*. Today, however, that pressure to cook fresh corn on the cob at once is considerably diminished as the varieties have changed. In addition to what is known as standard sweet corn, growers have bred what are known as Sugar-enhanced (se) sweet corn and Supersweet (SH2) sweet corn. The sugar content of the latter is higher than the other types and the crisp, tough skins on the kernels mean SH2 remains sweet a number of days after being harvested. As a result, SH2 sweet corn is popular among commercial growers since this stability means it can be stored and shipped successfully.

David Kinch, the acclaimed chef patron of Manresa, his three-Michelin-starred restaurant in California, has vivid memories of the sweet corn he ate as a boy growing up in Pennsylvania. 'It was one of those great summer treats,' he says warmly. 'The corn grew where my grandparents lived, which when we were kids was where we spent every weekend. My aunt and uncle grew corn themselves, so would pick and eat it. It had really, really small kernels, white in colour, you could almost eat it raw. It didn't need to cook for long and with sweet butter, it was pretty amazing.' He also remembers eating a Pennsylvania Dutch chicken and corn pie, made from roast or boiled chicken, the meat picked and mixed with fresh sweet corn and baked in a pie. 'You'd have a slice of pie – everything spilling out of the sides – and they would pour a glass of warm milk over it. It's pretty nineteenth century.'

While working as a young cook in France in the early 1980s, he was teased about American cuisine by the French chefs, Kinch tells me. '"What's a great American ingredient? What's a great American dish? Hamburger? Coca Cola?" I would talk about sweet corn and they'd laugh at me. "Corn, we feed it to our pigs, we don't eat it here", they said.' Kinch pauses reflectively. 'Of course, things have changed dramatically. Corn's a very trendy ingredient in France right now,' he observes wryly.

Kinch is a pioneer of the farm-to-table movement, working for several years exclusively with local growers Love Apple Farm and showcasing fresh produce on his menus. Despite his fondness for sweet corn, Kinch does not cook with it at Manresa. This is on principle, he explains. 'The corn today has been bred to be really super-sweet, because that is the defining characteristic that people are trying to capture in sweet corn. It is so sweet that it's more suitable for desserts – the sweetness is that intense. It is now so genetically modified that I tend to stay away from it.' He hankers after the sweet corn of his youth. 'The Mennonite communities where this corn used to grow were farming organically, some even biodynamically. It was part of their culture, part of their way of life; they were simple, God-fearing people taking care of the soil. The corn today doesn't come close to what I remember as a kid.'

WEEKS

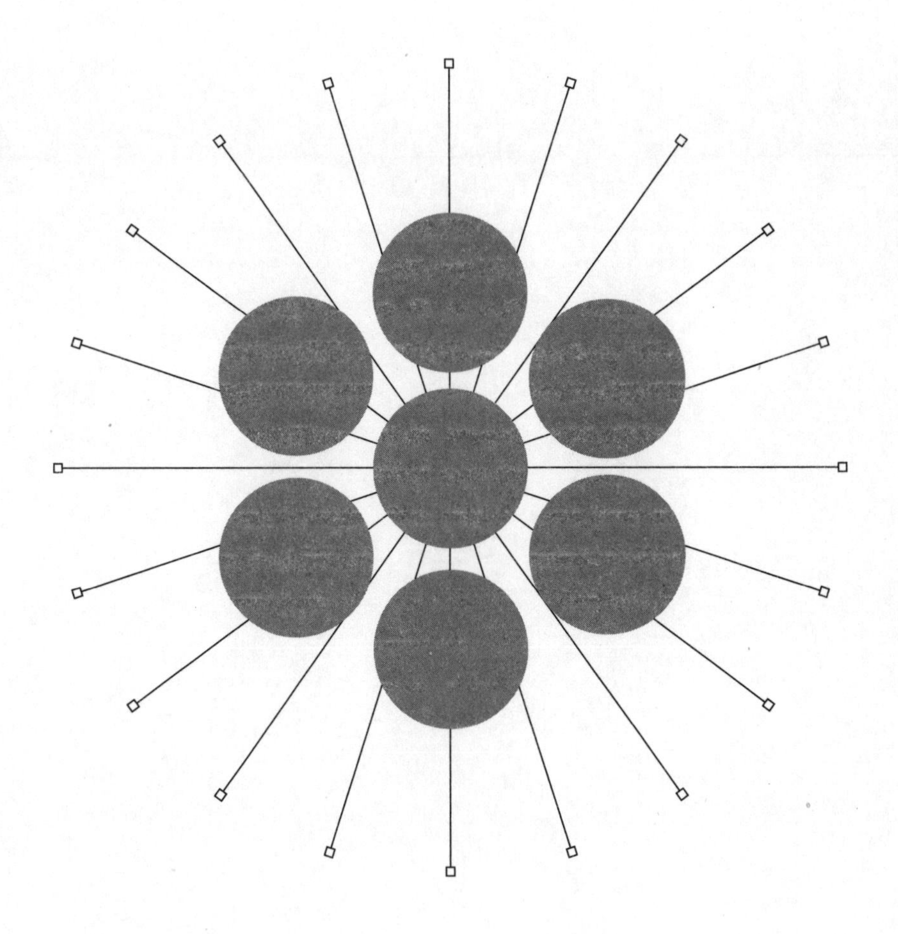

WEEKS

WEEKS

Weeks play their part in certain foods: the lives of poultry for our table are measured in weeks. Many of the cheeses we enjoy eating take weeks to create. These 7-day periods are used both by the cheese-makers when making the cheeses and by the cheesemongers, whose skill in affinage *(cheese-maturing) sees cheeses looked after and worked on (through techniques such as washing the rind) to create enjoyable, interesting cheeses that are ready to eat. The hanging of beef also takes weeks, contributing noticeably to the meat's texture and concentration of flavour. Craft beer, as with craft baking, involves substantial amounts of time being put back into the brewing process, in contrast to rapid, industrial brewing.*

THE MAKING OF CHEESE

For thousands of years, people have transformed precious, nutritious milk into cheese, changing it from a highly perishable liquid into a food which could be stored for longer, transported and traded. The practice of cheese-making is thought to have begun around 5,000 years ago in the Middle East and Central Asia, presumably in response to naturally occurring curdling in the hot climate. As with many of our preserving traditions, what began with an imperative, practical motive has developed into a craft of food appreciated in its own right. In an age of refrigeration, the need to preserve milk by transforming it into cheese is far less pressing, yet cheese continues to be made and enjoyed. Indeed, many countries around the world, such as the USA or Australia, have seen a thriving artisan cheese-making scene develop in recent years.

To walk into one of the world's great cheesemongers – Androuet in Paris, Fermier in Tokyo, Murray's Cheese in New York City, Neal's Yard Dairy in London – is to be reminded of the vitality of the cheese world. The variety of cheeses is extraordinary, as each counter will attest. All the more so when one considers that the starting point for each cheese is the same: milk. Over centuries of cheese-making, humans have found numerous ways to create a wide range of flavours and textures in cheese, but fundamentally cheese is made by curdling milk through the introduction of a coagulant, such as traditional animal rennet, causing the proteins within the milk to form what is called curd, from which liquid, known as whey, is expelled. (Whey, with its remaining proteins, can be re-curdled and used to make cheeses such as ricotta, whose name means 're-cooked', but was often traditionally fed to pigs.) From this fundamental starting point, however, the cheese-maker has a huge number of options as to how to create a cheese, depending on what results she or he is after. Cheeses are usually grouped into broad categories by type: soft, hard, washed rind, blue. In making these different styles of cheese, cheese-makers use a range of factors: handling the curds through draining and pressing them, bacterial cultures, heat, and – importantly – the passage of time. A fresh cheese, for example, is a simple type of soft cheese, made from heaping the curds in moulds, allowing it to drain simply under its own weight to create a delicate, moist texture, and is eaten just a few hours or a day or so after being made, before any rind has developed. The high moisture levels mean that they *need* to be eaten fresh, given that they spoil soon after making, as bacteria thrive in those moist conditions. Working and pressing the curd in order to extract moisture allows for the creation of hard cheeses, with their firmer, denser paste, cheeses that by their nature can be matured for longer and which develop a huge palette of flavours, among them nuttiness, fruitiness, umami and an acid tanginess, to name but a few.

Where cheeses require a longer period of time in order to develop their characteristic flavour and texture, part of the cheese-makers' skill rests in creating the optimum conditions for that cheese to be successful, the right atmosphere in which the vital processes that shape that cheese can take place. Blue cheeses, such as France's Roquefort or Italy's Gorgonzola, are created over a period of a few months by encouraging the growth within the cheese of blue moulds, which give the characteristic blue-green veining. Blue cheeses are not pressed; rather, the curds are loosely packed, left deliberately moist with space in between them in order to encourage the blueing. There is also a tradition among historic blue cheeses, including Britain's Stilton, of piercing the cheese at a few weeks of age to stimulate the growth of the blue mould.

The making of large hard cheeses requires a considerable investment of resources and time by its makers. Tying up substantial quantities of milk for several months before the cheeses are ready presents economic challenges. Switzerland's magnificent Emmentaler, for example, requires 1,000 litres of milk for one cheese, so was historically made by farmers pooling their milk through co-operatives. For a cheese to age well, it needs to be made correctly in the first place.

The complex, multifaceted role of time in cheese-making becomes apparent on a visit to Somerset, to the farm of Jamie Montgomery, the third generation of his family to make Cheddar here. One of a handful of Artisan Somerset Cheddar–makers recognized by Slow Food, Montgomery's farmhouse Cheddars – characterized by a dry texture and, often, an umami-rich savouriness – are admired around the world. In his huge storerooms, hundreds upon hundreds of Montgomery's carefully made Cheddars are aged for several months, batches tested and monitored throughout. 'Everything can go wrong,' says Montgomery with a short laugh, when asked about this maturing period, mentioning black mildew mould as one example of a risk that requires guarding against.

Although Montgomery is producing cheese that won't be eaten for several months, there is a daily rhythm to what he does, as the farm's raw milk is used to make cheese every day – Christmas Day included. The addition of traditional starter cultures – the batches of micro-organisms that cheese-makers use to start the process of changing milk into cheese – is used to create the cheese's characteristics. Another of the issues that

Montgomery has to combat in making his world-famous Cheddar is 'phage' (bacteriophage), a virus which, although not harmful to humans, attacks the lactic-acid-producing bacteria in the milk, so inhibiting the acid production and fermentation process needed to make cheese successfully. In order to prevent phage from developing, Montgomery uses a rotation of bacterial starter cultures. Each of these starters, however, affects the cheese slightly differently, meaning that each batch of Montgomery's Cheddar made on a particular day of the week has a distinctive set of characteristic flavours. Watching Montgomery at work sampling, tasting and assessing his Cheddars at 9 months with concentration and efficiency, I see for myself the intimate knowledge a cheese-maker has of his stock, especially when he works with the cheeses for so long. As he samples them, Montgomery is looking for certain traits to have developed in the intervening months according to their starters: 'It's like having seven children – if they behave the way you expect them to all is well . . .' Two starters, in particular, create the 'cream of the crop'. He discusses the behaviour of his cheeses as they mature with a distinctly parental affection. Having sampled one batch, a small smile appears on his face. 'This starter is the quiet child who sits in the corner – *doesn't* play a musical instrument – but this month it's blossomed,' he says with satisfaction.

Whereas the industrialized world of cheese-manufacturing is about creating consistency, farmhouse cheese-makers, while seeking to create a consistent level of quality, accept that variations in their cheeses from batch to batch are inevitable due to a complex host of factors, such as changes in the milk (depending on whether the dairy animals are feeding on grass or silage), the weather, the moisture in the air. The seasons, too, play their part. Goat's milk cheeses were historically made in spring, when goats had kidded and had ample milk. Among the great, seasonal traditions of the cheese world is the making of *alpage* mountain cheeses in Europe, which sees herds of dairy cattle in France and Switzerland taken up each year into the Alps during the summer to graze on the lush, flower-rich meadows there. The Alpine cheeses made during these months – such as Beaufort Chalet d'Alpage or Gruyère d'Alpage – are noted for their fragrant, floral flavours which come through the milk. These variations from batch to batch among artisan cheeses make cheeses fascinating for those who work with them – producing them, maturing them and selling them – and for those who eat them. The best cheesemongers give tastings of their precious stock, appreciating that it is important for their customers to try a particular cheese, even if it's of a type familiar to them, made by a maker they like, before they buy it. Buying cheese from a good cheesemonger, having tasted and discussed it, is a leisurely experience, not to be rushed, a shopping trip with an element of exploration to it.

The industrialization of cheese-making during the nineteenth and twentieth centuries resulted in the considerable speeding up of the process. Both World War I and World War II contributed to the decline of traditional cheese-making, owing to the loss of farming manpower and, during World War II, the impact of the centralized collection of milk by the Ministry of Food. In his seminal 1982 work, *The Great British Cheese Book* – a clarion call to save Britain's cheese heritage – cheesemonger Patrick Rance eloquently recorded the blighting effect. 'In 1948, 44 farms in Cheshire produced their own cheese, compared with 405 in 1939; 29 against 202 in Lancashire and 61 against 514 in the South West.' Cheeses, rather than being made on farms by families, as had been the custom for centuries, were now predominantly produced by rote in factories, often made in large blocks (which never develop a rind) which are sliced and packaged, rather than individual cheeses. But, inspired partly by figures such as Patrick Rance and Randolph Hodgson of Neal's Yard Dairy, Britain has seen a revival in the craft cheese-making movement, with many using historic methods to make entirely new, original cheeses. Other farmhouse producers have carried on maintaining traditions passed down within their families.

On his family's dairy farm in Lancashire, Lancashire cheese-making has been part of third-generation cheese-maker Graham Kirkham's life ever since he was a young lad. Relaxed and jovial in his speech, Kirkham is also fiercely committed to maintaining traditions. At every step of the way, great care is taken. The raw milk he uses comes from his own herd of Friesian Holsteins, with the focus of his dairy management being on the quality of milk rather than the quantity. He begins by mixing milk from the previous evening's milking with fresh morning's milk, to which he adds first a small quantity of starter culture, then, a little later, the natural rennet to set the milk, 'like a big blancmange'. In order to release the moisture (the whey), the curds are cut into 10cm cubes using hand cutters, a gentler method than mechanized overhead cutters. These warm curds are cut, moved and stacked for an hour and a half, in order, Kirkham explains, to help them grow in acidity, strength and elasticity. The curds are then torn into pieces, pressed and left overnight, before being cut again to drain out moisture. There follows a long, intricate process of pressing, breaking, cutting and resting, milling (cutting into small pieces) and salting the curds. 'The reason I'm doing all this faffing around,' observes Kirkham matter-of-factly, 'is because we use a very, very small amount of starter culture.' The more starter culture a cheese-maker uses, the more acidic the cheese becomes, resulting in the moisture being forced out quickly through the reaction of the starter culture with the milk. The acidity, however, affects the flavour of the cheese, so Kirkham opts for a slower, more laborious method of removing the whey from the curd.

Kirkham then puts milled curds from 3 days into moulds. This use of curds from different days is a defining characteristic of a traditional Lancashire cheese. As he points out, dairy farmers in the past wouldn't necessarily have had enough curd from one day's milking to mill and fill a 50lb (22kg) mould, so they worked using curd from 2 or 3 days' production. In addition to maintaining the tradition, there is good reason for Kirkham in mixing the curds in this way: 'It definitely develops more flavour. Your third-day curd has changed in texture, body and flavours; it mellows and goes quite nutty.' Lancashire cheese made with a single day's curd, in contrast, he feels is 'dry and mealy'.

The curds are pressed overnight to form the traditional truckle shape, wrapped in muslin, pressed again and left to dry out for 2 days. Clarified butter is then spread over the muslin, to moisten it, and the cheeses are matured on Kirkham's farm, usually until 3 or 4 months old. In contrast to the way Kirkham makes his cheese – over days – industrialized Lancashire cheese is made from curds to moulding in hours. At 3 or 4 months, Graham's preferred age, Kirkham's Lancashire is characterized both by its gentle dairy flavour and by its unique crumbly texture. Although it is considered a hard cheese, one can break the paste simply using one's fingers, without having to cut it with a knife. Lancashire cheeses were made as far back as the thirteenth century. Today, Kirkham's, made slowly, carefully, skilfully by Graham Kirkham from his own milk on his own Lancashire farm, using traditional methods, is the only raw-milk farmhouse Lancashire still being made.

CRAFTING BEER

The brewing of beer from fermented grain has been practiced for millennia; Sumerian clay tablets dating from 3000 BCE offer evidence that beer was made then. The knowledge of how to brew this alcoholic drink is assumed to have travelled from the Middle East to Europe. Beer-making was taken up with enthusiasm by northern countries, whose cold climates precluded grapes being grown and turned into wine. The Roman historian Pliny the Elder (AD 23–79) recorded that beer was drunk by the Saxons, Celts, Norse and German tribes. Preserved tablets at Vindolanda, the Roman fort just south of Hadrian's Wall, mention beer. By the Middle Ages beer was widely drunk, and not just because it was alcoholic. Because it was boiled, beer was safer to drink than water; it was rich in carbohydrates – so an important source of calories – and had the advantage that it could be stored safely without the risk of rotting. Early beers were flavoured with herbs and spices. The use of hops – the flowers of the hop plant – as a customary flavouring for beer came about later. An act prohibiting the adulteration of hops was passed in England in 1603, and beer made with hops became increasingly popular. The eighteenth and nineteenth centuries brought innovations in British brewing, including the development of porter, stout and India Pale Ale (IPA).

Britain's long history of brewing and drinking beer – and the consequent widespread presence of pubs on our high streets – came under threat during the 1960s. A wave of takeovers and mergers of small regional breweries saw the creation of six huge companies. The biggest of these was Bass Charrington, which owned over half the country's pubs. The focus of these large companies was to promote mass-manufactured beer, rather than the traditional, flavoursome, slow-fermented beers enjoyed in the past. 'Real ale,' explains beer journalist Roger Protz, 'like all the best things in life, is made carefully and slowly. It reaches fruition in its cask in the pub cellar, where it undergoes a secondary fermentation.' In contrast to this, 'industrial beer' is filtered and pasteurized. Cask-conditioned beer furthermore requires a wait of at least 24 hours for the beer to settle and become clear. The big breweries wanted to deliver kegs of beer that could be used in the bar at once. This move towards uniform, fizzy, bland, mass-produced beer saw a backlash from British beer-lovers. In 1971 the Campaign for Real Ale was set up, an organization committed to championing traditional beer. The campaign struck a chord, with the first CAMRA Beer Festival in 1974 attracting a huge attendance. This ongoing work by CAMRA to preserve historic ways of brewing and storing beer laid an important foundation for what was to come.

The decade that saw British beer-lovers fighting to preserve its heritage also saw the beginnings of the rise of a craft beer movement in the USA. A trigger for this was legislation passed in 1978, legalizing the home production of a small amount of beer for personal consumption. The beer scene in America was dominated by large, industrial brewers; home-brewing offered beer enthusiasts the chance to explore beer's potential. The interest in beer saw the rise of microbreweries. Whereas in 1978 the USA had only forty-two breweries, by 2012 it had over 2,750. The craft beer movement has seen a similar expansion in breweries in Britain. 'It's incredibly exciting,' says Protz, with palpable relish. 'The first *Good Beer Guide* in 1972 listed forty-four working breweries. There are now 1,700 working breweries in this country.'

In the peaceful, rural setting of the Welbeck Estate in Nottinghamshire, Welbeck Abbey Brewery, founded in 2011, is among these new breweries. The brewery's general manager, Claire Monk, is proud of what they do. 'We only use malted barley, hops, yeast and water. We have a strict rule that we don't put any rubbish in our beer,' she tells me matter-of-factly.

Monk studied microbiology and biochemistry at university and her scientific approach to the business of brewing becomes apparent as she talks. Various factors contribute to the final taste and texture of beer. Brewer's yeast – which turns the sugars into alcohol and carbon dioxide – contributes around 20% of the flavour. Different yeasts are used to create different beer styles. Welbeck Abbey Brewery produces predominantly traditional real ale, a beer that Monk herself enjoys drinking and for which there is demand: 'We're on a historic, landed estate that would have made real ale since the year dot, so it fits in nicely with where we are.'

Real ale is warm-fermented, using a traditional, British style of 'top-fermenting yeast' (one in which the main body of yeast does its work at the top of the liquid) that gives a particular flavour profile. 'When you package the beer into casks, a little bit of that yeast goes in and keeps on working and that's what makes it a live real ale.' Lager, in contrast, is cold-fermented, using a bottom-dwelling yeast, then filtered or pasteurized, with the final product no longer live. The yeast used for the majority of the beers at Welbeck Abbey is the brewery's own yeast, nurtured and looked after by Monk and her team. 'When we brew we recover a lot more yeast than we put in in the first place. Unlike baking, where you kill off yeast, we are constantly propagating it. Historically, bakeries used to be next to breweries for that reason. The people at the bakehouse here on the estate often pop over and get yeast from us.' The water used for brewing is also their own, sourced from the estate's borehole.

When it comes to the malted barley, Monk explains that using combinations of roasted and crystallized barley allows brewers to change the flavours in their beers. Crystallized malted barley is barley heated so that the natural sugars within it caramelize. 'It's like making caramel at home; the more you cook it the darker it goes. I can buy a range, from pale crystal malt, which is sweet and biscuity, all the way to extra-dark crystal malt, which has a burnt-toffee flavour.'

The usual process for making real ale at Welbeck Abbey takes 5 days. It begins with boiling the wort (the sweet, sugary liquid created by soaking the malt) with whole flower hops, giving the mixture a characteristic bitter flavour. Hops are also added at the end of the boiling, to give aroma, and left for half an hour. 'We say it's like steeping a tea-bag!' The next stage is to ferment the beer, which takes around 3 days, depending on the strength of the particular beer being made (a factor which varies with the amount of natural sugars being fermented). On the quest for consistency this is done at a strictly controlled 71°F. If you ferment above 77°F, 'weird and wonderful' things can happen. 'As a scientist, unknowns and variables stress me out, so we control as much as we can.' In order to create Welbeck Abbey's Henrietta beer, the fermented beer is then transferred to an enclosed tank. Here hops are added, the temperature is lowered and the beer is left for up to a week to cold-infuse, a process also known as dry-hopping. 'The reason we do this is that a cold infusion gets a different dimension of the hop character. We are using a very subtle German hop and using it in both the boil and the cold infusion means you end up with quite a complex beer, that is floral, grassy and really fresh.'

Once the real ale has been made, it goes into barrels where it requires at least a week to develop, a process called conditioning. This important stage takes place either at the brewery or in pub cellars. During this time the yeast works on the sugar in the beer, converting it into carbon dioxide. 'That means when you have a pint of beer in the pub, it's not flat, although it's not fizzy like lager. It's got a rich texture to it, from very small carbon dioxide bubbles.' Before conditioning, Monk tells me, a beer might be overly bitter and harsh. 'The conditioning smooths it out, giving a more mellow bitterness. It also creates texture, giving a much better mouth-feel.' From the brewery's point of view, working with pub landlords who care about the beer they sell and know how to correctly condition it is essential; lack of attention at this final stage can undo all the previous careful work.

It is not just the number of breweries that heartens Protz about Britain's current beer scene, it is the range of beers now being made and the creativity being shown by the new generation of brewers. 'Back in the 1970s breweries produced two types of beer; mild and bitter. Look at the choice now! Genuine Indian Pale Ale, proper porters and stouts have been brought back, people are ageing beers in wood. The whole way in which beer is being made and presented is different – and it's wonderful! Some of the new brewers here and in the USA are breaking down boundaries and reaching out to a younger audience with exciting flavours, and that's really important for the future of beer.'

IN DEFENCE OF HANGING

Even though it is a mild spring day outside, I regret not having worn warmer clothes. I am standing in a large cold room – a space living up to this description – looking at row upon row of hanging beef carcasses. Despite this proximity to so much dead meat, the smell in the air is not one of decay or blood or mould; instead it is a noticeably savoury smell – far from unappetizing – one which brings the word 'umami' to mind and makes me feel distinctly carnivorous.

The human love of meat has ensured that, as a species, we have gone to considerable trouble to eat it, from hunting animals to rearing them for the table. With eating fresh meat, however, comes danger – so nutritious is meat that bacteria love it, carrying the very real risk of food-borne infections. There are many historic meat taboos, linked to the perils of consuming meat which might have become infected with parasites. Much ingenuity has been spent in ensuring that the meat we eat is safe to consume, fighting the effects of time, the natural, rapid deterioration fresh meat is subject to and the danger that brings. Canning, refrigeration and freezing are all methods used to make sure that the meat we eat is safe. In many countries nowadays, from the moment an animal is slaughtered to its final end as a piece of meat on sale, there is a continuous carefully monitored cold chain, ensuring that it stays fresh. This cold room, however, is a deliberate pause on that journey from abattoir to plate – in effect a place where a skilled butcher is carefully harnessing the process of decay to create meat that will be both tender and more flavourful. With its constant supply of cold air pumped around the space, this room is where meat is hung, a traditional process used to enhance its taste and texture.

As I look around, the word 'hanging' takes on a clarity of meaning. Carcasses are indeed suspended – in appropriate cool and dry conditions – and left to age for several days. Today, in our increasingly squeamish world, the terms 'dry-ageing' or 'maturing' are often used to describe the same process. Hanging meat, which relies on keeping the meat cool, is a process traditionally favoured in countries with cooler climates, not tropical ones, where meat perishes quickly due to the high temperatures and humidity. Hanging meat is effective because once the animal has been slaughtered the muscle enzymes within it begin attacking cell molecules, breaking down proteins into savoury amino acids. This process releases tasty by-products that contribute to the distinctive savoury flavour of aged meat. The enzyme activity during the hanging process also affects the texture by breaking down the protein structure and weakening the collagen in connective tissue, resulting in meat which is more tender when cooked. In addition, the meat dries out during this period, which also ensures a soft texture when cooked. In contrast, the water present in fresh meat, which hasn't been hung, expands during the cooking process, then leaches out afterwards, resulting in dry food.

It is easy to experience for yourself the difference that hanging meat makes to both its flavour and its texture. Steak is historically a luxurious cut of meat, with a whole beef carcass yielding only a few of these prime cuts. Buy a cheap steak from a supermarket, however, with its red-pink colour and bright white fat, and when you eat it, it is hard to understand why this meat would be considered a treat. The flesh has a moist, spongy texture with a leathery finish when chewed, and there is a noticeable lack of tastiness; at best it is insipid in flavour, at worst there is an unpleasant sourness. In contrast, steak bought from a good butcher, although it comes at a noticeably higher cost, reminds you why steak has its reputation as a treat. The dark-coloured meat, with its yellowed fat, is at once tender, yet textured rather than watery. The flavour is rich, savoury and profoundly satisfying in a primal way. The difference is noticeable – a revelation. One of the prime reasons why meat, especially beef, will taste better when sourced from a good butcher's shop is because of the work that butcher will have done to improve the meat by hanging it.

In Britain there was a long, rich heritage of hanging meat, classically beef (as the size of the carcass allows it to be hung for several days) and game. Victorian photographs of butcher's shops – carefully constructed with marble counters and tiled walls to keep the meat cool – are adorned with hanging carcasses. The rise of supermarkets as the convenient one-stop place in which we shop for food, including meat, has seen the notable decline of high-street butchers, and with their vanishing has come a loss in the art of hanging meat. Hanging meat requires time, specialist equipment, space and expertise. As a carcass hangs in a cold, dry atmosphere, it loses moisture and therefore weight, between 10% and 20% during the hanging period. Furthermore, the surface of the meat dries out and becomes mouldy, so this needs to be expertly trimmed, resulting in further loss. In short, hanging meat brings with it a cost. The commercial imperative to maximize the return on each carcass ensures that most of the meat sold in supermarkets has not been hung. In the quest for the high-end market, some supermarkets will age small cuts of meat, so that they can then offer a range of prime cuts such as steak with labels such as 'dry-aged for 30 days'. The term 'aged' when seen on a meat label is deceptive, as rather than hanging or dry-ageing this generally implies a process called 'wet-ageing', where meat is wrapped in plastic and aged for days or weeks. The plastic wrapping ensures that meat keeps its moisture – so no shrinkage occurs – and there is some enzyme action which enhances flavour and texture, though to a far lesser extent than with traditional dry-ageing.

Nathan Mills at The Butchery is one of the new generation of young butchers making a name for himself in the United Kingdom's competitive food scene, supplying meat to talented chefs at restaurants such as the Dairy and Pitt Cue which prize quality, provenance and local food, as well as selling to retail customers. Mills is noted for his flavourful meat, notably his beef, which he hangs for 55 days, considerably longer than most butchers, up to 21 days being the norm. Ironically, as a third-generation Australian butcher, when Mills came to work in the UK he was amazed at the way in which beef was hung. 'I was at Smithfield when I had my first eye-opening moment, when I walked into a cold room and saw a piece of beef that was furry [i.e. mouldy]! That wouldn't have passed in Australia, but it's a different culture. Here it's part of the tradition.'

The huge cold room I'm standing in is within the Butchery's headquarters, located under a railway arch in Bermondsey, south London. At £20,000 it represents a considerable investment by Mills in his fledgling business, as, too, does the meat stored in it. Having this large, refrigerated walk-in space allows him to add value to the carcasses he buys by hanging the meat himself until it reaches the stage he wants. The fresh carcasses are brought in from the abattoirs and hung at the coldest end of the room to bring the temperature of the meat down as fast as possible. As they age, they are then moved down the room. During this time they acquire a 'cap' of white mould. Not only must the room be cold, it requires air-flow, provided by large and noisy fans which whir away. Warmth and humidity are the enemies here, with a big delivery of 3–4 tons of warm, fresh meat needing to be cooled as quickly as possible. To that end, there is regular monitoring of both the temperature of the atmosphere – the air flowing out is about minus 37°F – and the meat itself, with the carcasses regularly probed, Mills looking for 32°F at the cooler end and up to 34°F at the other.

For Mills, the starting point of creating meat which has been well hung is the quality of the carcass itself. Ageing will not improve bad beef, he tells me. For Mills, choosing the right breeds to hang is important. Rather than buying his meat from third parties at Smithfield, Mills sources his meat from farmers of native and rare-breed cattle, working directly with over thirty farms to ensure his supply. The reason he's chosen native breeds, he explains, is that their meat 'lends itself to hanging', as the process enhances the depth of flavour these slow-raised animals possess as well as creating tender meat.

'Fat cover' and 'marbling' (the spread of fat through the meat) are terms that come up quickly once you begin discussing hanging beef carcasses with a butcher. In order to hang well, and for optimum yield from the butcher's point of view, a beef carcass needs a good 'covering' of subcutaneous fat. Beef cattle are stripped of their hides for the leather industry and this skinning process 'makes or breaks the carcass'. Mills points out a large tear in one of the beef carcasses hanging up. 'That's not great for me in the ageing process, as the fat would have covered that and protected it.' He pats another carcass affectionately. 'Here they've done an amazing job. They've kept all the fat on, which is good for me. This will hang really well.'

Too much fat, however, is not a plus point, as the customer

wants to eat meat, not fat. Native animals feed off UK grass and have to be grown 'long and slow', with Mills looking for animals at 30 months. If they are fed grain and silage to speed up their weight gain, they inconveniently transform that into fat rather than muscle (i.e. meat). In contrast, modern continental breeds which are fed a lot of grain and silage, put on weight quite quickly, but not an excessive amount of fat, so they can be slaughtered at 18 months. Mills, however, feels that the time required to bring on native breeds is reflected in their superior flavour. Dexter cattle are his best-sellers, with customers coming in specifically for Dexter beef. 'It's a beef connoisseur's beef. They age really well. Every chef I send Dexter to asks for more.'

Mills's commitment to the process of producing dry-aged meat is impressive, from the time-consuming sourcing of native breeds, through the vigilant care taken during the hanging, and the final, expert trimming of the meat – sharp blades gliding through the flesh with speed and ease. Why take the time, effort and money to hang meat this way? I ask him. 'Because you get flavour and tenderness.' He pauses, then says simply, 'Look, basically, it's like making a stock. You reduce the moisture, you intensify the flavour.'

PRODUCING POULTRY

Given how cheap industrially produced, intensively reared chicken is, it is not surprising to learn that the birds' lives are short. Livestock farmer Peter Greig of Pipers Farm worked in his father's industrial chicken farm and has seen for himself the way the poultry industry has gone. 'When my dad first brought over industrial chickens from America, it would take about 58 days to produce a four-pound chicken. Before he died, that had been reduced to about 35 days. Now, they can produce that weight of chicken in about 28 days. There is this constant process of honing the genetics and the feeding regime to make things happen faster and faster.'

The birth of their sons led to an epiphany for Peter and his wife, Henri. Understanding both the living conditions these intensively reared chickens were enduring and the amounts of antibiotics routinely used to keep the poultry alive, they realized that they did not want to feed this type of chicken to their children. Greig left his father's business and he and his wife set up Pipers Farm near Cullompton in Devon in 1989 to produce poultry and meat that they would happily give to their boys. 'Over all the years, most of our customers have been young families,' Greig reflects. 'Becoming a parent clearly is a moment when many people stop and think about what they're eating.'

Greig's chickens, bred from the farm's own flocks, live a free-range existence from the age of 3½ weeks on the hillside above the farm. This outdoor life, with the birds importantly exposed to ultra-violet light, helps them build up a natural immune system. The resulting birds have noticeably strong, firm bones, filled with nutritious marrow, from which one can create an excellent jellied stock. They are killed at 12–13 weeks (rather than 35 days for the most intensive housed systems or 10 weeks for mainstream free-range), then dry-plucked and hung for around a week, the way Greig's great-grandfather did. 'Hanging is about time and temperature,' he tells me. 'If the weather is cooler it might be for 12 days.' Taking the trouble to hang chickens is unusual these days, but Greig feels that this natural maturation process is an important part of developing the flavour. 'What we are looking for in all our meat and poultry is texture and also a lingering depth of flavour that goes right to the back of your palate.'

When Pipers Farm was set up, the ideas behind it were radical and Greig often felt like a lone voice. These days he is heartened by the interest in good food in Britain. He tells me, with real excitement, how he and Henri cooked for sixty people a few days ago at the reopened cider mill down the road. 'It felt like a dream. I remember thirty years ago seeing French farmers sitting at long tables, eating and drinking what they had produced and spending 3 hours over it. Everything about that meal – people from the local community producing food, respecting it, enjoying it – felt like a significant milestone on the journey.'

Just as chicken farming has become intensified, so, too, has turkey farming. The turkey is now Britain's traditional Christmas bird, long ago overtaking the goose in popularity, perhaps inspired by its festive role as a Thanksgiving centrepiece in America, where the turkey originated. Despite the bird's celebratory role in our lives, the realities of most turkey farming reflect the industrialized nature of the business. Since the 1960s, the modern commercial turkey has been bred to put on a huge amount of weight in a short space of time, significantly reducing its allocated lifespan. In the 1970s a turkey would have taken 20 weeks to reach the desired weight. An industrially raised turkey now takes only 10 weeks to reach 5kg (11lb), at which age it is slaughtered. In effect, the birds are killed when they have reached the desired weight, but before their bodies have had time to mature and develop strong bones or build up fat that would make the meat tender and juicy. Once slaughtered, the birds are processed speedily. They are dipped into a scald tank of hot water, passed through a mechanical plucking machine that removes the loosened feathers, then have their internal organs removed and are washed and chilled.

Self-styled 'maverick' Paul Kelly of Kelly Bronze Turkeys champions an alternative approach to turkey farming. Turkey farming is a family affair; his father bred white turkeys, which were dry-plucked and hung and sold to butchers. The early 1980s, however, saw their business threatened, with the decline of butchers and the increase in cheap, fresh turkeys offered by the supermarkets. Kelly realized that the only way to compete was to offer a different, high-end turkey. This once traditional bronze turkey breed had fallen out of favour due to its dark feather stubs, which, visible after plucking, were considered unsightly. Daringly, Kelly set out to revive the breed, rearing the birds free-range. The lifespan of a Kelly Bronze turkey is 6 months. After slaughter, the birds are dry-plucked. This process keeps the skin intact and so permits hanging in order to create further flavour in the birds. See a Kelly Bronze turkey side by side with a conventional, industrially reared one and the difference in shape is striking. The former is noticeably more athletic and slender, lacking the huge breasts of the latter. The Kelly Bronze turkey meat is dense, moist and flavourful.

Kelly, having taken his family business in this bold direction, vividly remembers how hard it was in those early days to find butchers to stock his upmarket poultry. An endorsement from influential food writer Delia Smith in her best-selling Christmas cookbook proved to be a turning point, and Kelly Bronze turkeys today enjoy accolades from leading chefs. Kelly has since pioneered the rearing of free-range turkeys in woodlands, their original habitat, rather than offering housing in sheds. A visit to his farm in Essex gave me a chance to see the turkeys in this setting. Within minutes of standing still and waiting, Kelly and I were surrounded by a huge flock of bright-eyed, inquisitive, handsome birds, a credit to his husbandry.

Ever one to relish a challenge, Kelly's latest project is to take his Kelly Bronze turkeys to America, where he has recently set up a farm. Turkeys in the US, he explains, 'are grown as quickly as they can be – not a culinary delight, to say the least. The tradition of dry-plucking the birds, we call it New York dressed, yet over in America there isn't one single person dry-plucking and hanging turkeys, so I'm taking the New York dressed turkey back to America.'

THE MATURING OF CHEESE

Artisan cheeses require nurturing. The hard work that goes into them does not stop once they have been made. While mass-produced, plastic-wrapped block cheeses are simple to keep, artisan cheeses are distinctly more demanding. They need to be stored in the appropriate conditions and cared for until ready. The French have a word for it: *affinage*, meaning 'the ripening of cheese', bringing cheeses on until they reach the right moment to be enjoyed. The reason why we have specialist cheese shops – cheesemongers – is because fine cheeses require much patient, knowledgeable care. Once out of its maker's hands, what was a well-made cheese can easily be spoilt. Drying out, cracking, sweating, the development of mould on cut surfaces, and plastic tainting are all possible if cheese is not looked after properly. Equally, though, an experienced cheesemonger can do much to enhance the cheeses being sold. When buying in young cheeses, a cheesemonger can mature them further to the point where he or she deems their taste and texture is at its best. Carrying out this process correctly requires knowledge, attention and patience.

The world of artisan cheese is one filled with variables. Batches of cheese, even from the same cheese-maker, vary depending on factors such as the feed the dairy animals were given, the quality of the milk, the weather while the cheese was made and ripened. Good cheesemongers relish the interest these differences offer. They prefer working with a complex, variable food to one that is homogenously consistent.

Visit Neal's Yard Dairy's flagship store by Borough Market and you will be struck both by the impressive display of British and Irish cheeses and by the helpful, friendly service. Behind the scenes, however, a huge amount of unseen work also takes place. Catching the first tube train of the day at 5.15 a.m. on a spring morning gives me an inkling of the effort gone to by Neal's Yard Dairy's staff. I am on my way to meet their buyer Bronwen Percival outside the Borough store at 6 a.m., accompanying her on her monthly cheese selection visit by car to Cheddar-makers in the West Country. These regular journeys to cheese-makers throughout Britain are to assess and select the hard cheeses, cheeses that once made are stored for months at the cheese-makers until ready to sell. These trips, therefore, offer a valuable chance for Bronwen and her colleagues to monitor the progress of the cheeses with their maker as they mature.

Our first stop is Manor Farm, North Cadbury, where Jamie Montgomery creates his world-renowned Cheddars. Standing in Manor Farm's huge, dark, cool storage room, I watch with interest as Montgomery methodically works his way down rows of cloth-wrapped cheeses, using a cheese iron to remove a core sample from his Cheddars, which we then taste. Both he and Percival are focused on the task at hand, assessing the cheeses, meticulously noting how each batch has developed since last tasted. Short interchanges regarding flavour, texture, expectations, potential take place; it is a thoughtful process. 'There's a savoury sparkle there,' Percival comments approvingly after one sampling. The ability to taste a hard cheese when it is young and be able to predict how it will develop over several months is one that takes time – and much sampling en route – to acquire. Percival recalls how when she first accompanied Randolph Hodgson, owner of Neal's Yard Dairy, on selection visits, he and Jamie Montgomery would taste a Cheddar and predict it would be great, whereas Percival wouldn't like it at all. 'Randolph said no, that's a cheese that's just going to take a long time to come round. It's so dry that right now it tastes muted and uninteresting, but if you keep it for 10 or 12 months it's going to be incredible. And I noted down the batches in my book and, lo and behold, when the time came to sell, the cheese was amazing!' Part of Percival's work here today is to allocate which parts of Neal's Yard Dairy's business to send these selected Cheddars to: retail, wholesale or for export to the US. There is an age element to her choices. The younger 12–13-month cheeses are exported to America, as shipping will take 4–5 weeks and the expectation is that they will be opened and sold. The oldest cheeses, up to 20 months, are selected for the shops. For these, Percival uses her knowledge and experience to assess how these large, cloth-wrapped cheeses will develop over the remaining months to get to the desired flavour profile that Neal's Yard Dairy looks for.

The work that Neal's Yard Dairy puts into caring for cheeses takes place at their maturing rooms, atmospherically housed under railway arches in south-east London. Here cheeses are stored for a number of weeks before being sold in the shops; this period often has a transformative element to it. The staff use the six rooms, each with their own different temperatures and humidity levels, to affect the development of the cheeses, giving, for example, soft young cheeses (which arrive at ages ranging from a few days old to a fortnight) time to grow a rind. When I arrive at 6.30 a.m., Joe Bennett of Highfields Farm Dairy, having driven down that morning from Staffordshire and delivered his weekly batch of soft goat's cheeses, is enjoying a cup of coffee and discussing how his recent cheeses are developing. This collaborative way of working, patterned by regular visits and feedback, has been very useful in helping him improve his cheeses, Bennett tells me. The weekly visits – giving him the chance to see how different batches of his cheeses have developed over the intervening days – offer useful insights.

'Working with the soft cheeses is intensive but for a short period of time. It's really interesting how they change so much from when they come in to when they go out,' Sarah Stewart tells me with tangible enthusiasm as she shows me round the different rooms, their environments ranging from chilly and dry to warm and moist. Encouraging the right sort of mould, depending on the type of cheese in question, is a key part of the work taking place here. It is a process that requires both the right conditions and time for the mould to grow. The 4-day-old Innes Logs delivered by Bennett will first be taken to the drying room (57°F, 60–80% humidity) and kept there for at least 24 hours in order to lose moisture, stabilize the rind and become firm enough to handle. They are then moved into the warmer, moister conditions required to encourage the development of *Geotrichum* mould on the rind of the cheese. Once the *Geotrichum* is properly established, the cheeses are moved to a colder, slightly drier room (50°F, 90% humidity) to control the development, and usually sold when 2–4 weeks old. Over this period as the *Geotrichum* grows it forms a characteristic wrinkled layer. The delicacy of its rind and the way in which it becomes an integrated part of the cheese is much appreciated by cheese aficionados.

Another mould that Sarah and her colleagues seek to actively cultivate on some of their cheeses is *Bacterium linens*, the mould that gives washed rind cheeses, such as Gubbeen, their characteristic orange-red colour, sticky rind, flavour and aroma. Here at the maturing rooms, much patient washing of whole cheeses over weeks takes place in order to transform them into washed rind cheeses. The washing removes the dominant white *Penicillium* mould, creating the less acidic, moist environment preferred by *Bacterium linens.* I watch as a batch of young Riseley cheeses made from sheep's milk, at the moment firm and easy to handle, are swiftly wiped with a brine wash. Now they are pale in colour, but over 4 weeks, with regular washing, the rind will take on a pinky-orange tone and the cheeses will soften and develop a noticeably savoury flavour. The conditions in the 'Stilton hastening' room – kept at 54–55°F with a humidity of around 96% – are also used to encourage the growth of *B. linens.* Part of the skill exercised by the team here is in using the different rooms with their varied temperatures to speed up or slow down stages of development as required. The colder spaces are used to act as 'brakes' on the process. A cheese might spend all its time in the maturing rooms in just one room or go into three or four rooms, depending on what is required. The good thing about working with soft cheeses, Stewart points out, is that one is able to quickly get 'feedback' to changes within a day or so, unlike hard cheeses where the consequences of a change become apparent only weeks or months later.

I am struck by how much incessant care and attention the cheeses here require; there is a patient, nurturing aspect to the work. Even the large hard cheeses, which are being stored here rather than developed, need turning once a week in order to keep the moisture content within evenly spread, rather than dropping to the bottom. There is a quiet, observant vigilance to Stewart's progress through the rooms, watchful for the formation of unwanted moulds, such as the expressively named 'cat's fur' mould. Stewart, it is evident, gets huge satisfaction from the constant, intricate cycle of cheese-maturing. 'When I come in after the weekend,' she tells me, 'the first thing I do is go through and taste everything and things will have changed immensely since I left on Friday. It's really interesting.'

After my visit to the maturing rooms, I walk for 20 minutes or so – passing the Shard en route – until I reach Neal's Yard Dairy's shop at Borough Market, a large, high-ceilinged space, well staffed in order to deal with the constant stream of customers. Here for sale are the cheeses that the maturing rooms work on, bringing them to the 'window' of time where they judge them to be at their best. I can see and sample the results of the washing team's patient work: dainty St Jude cheeses, for example, sit alongside their washed rind version, St Cera. On the counter I spot Graham Kirkham's farmhouse Lancashire cheese. It is sold here not just at the usual 3 months, but also in a matured version, several months older, Neal's Yard Dairy having experimented with ageing it in the Borough Market shop. Intrigued, I buy a piece of each. The young Lancashire is soft and moist, with a delicate buttery flavour. The 9-month Lancashire is noticeably drier and firmer, although retaining its traditional crumbliness. The flavour in the older version has changed, developing a salty bite, a kick to it, with a lingering savoury finish. Both cheeses are fine examples of each age – a tangible, edible testament to thoughtful *affinage*.

SUGARING RUSH

The books we read and love as children lodge in our memories in a very particular way. I first read Laura Ingalls Wilder's *Little House* series as an eight-year-old child living in Singapore, a tropical world away from the rural American childhood she described. In her books Wilder conveyed a sense of how hard-won food was – how much work went into its growing, gathering, hunting and cooking – which impressed me. I didn't know what wintergreen berries tasted like (indeed I still don't) or what johnnycake was, but I used my imagination.

Of all the plentiful food encountered in those pages, however, it was the descriptions of maple syrup in *Little House in the Big Woods* that struck me most. Pa explains to Laura how all winter Grandpa has made wooden buckets and whittled dozens of little wooden troughs. With the arrival of the first warm weather he fits these troughs into holes bored into maple trees. As the maple sap, 'the blood of the tree', rises in each tree it runs down these troughs to be collected in the buckets placed below. This sweet sap is then boiled down until it turns into syrup and crystallises into sugar. In the next chapter, family and friends gather together for a dance at Grandpa's house. The evening ends with a feast, the highlight being Grandma ladling hot 'waxing' maple syrup onto platefuls of freshly gathered, clean, cold snow – a wonderful image which has stayed with me ever since. To this day, maple syrup retains a sense for me of being a special treat.

Maple syrup is produced in northeastern North America; Quebec in Canada is the largest producer overall, while within the United States Vermont produces the most syrup.

It is the sap from three varieties of maple in particular: the aptly named sugar maple (*Acer saccharum*), the black maple (*Acer nigrum*) and the red maple (*Acer rubrum*) – which is most widely used. A cold climate is essential to the creation of the sweet sap, allowing starches stored by the tree to be converted into sugar over the winter months. The arrival of Spring sees the rising of sap up into the tree, with the sugar, along with nutrients, sent from the roots to feed the tree's new growth. The process of tapping the sweet spring sap from the trees and cooking it down to form a syrup or until it crystallised into a sugar was practiced by the indigenous peoples of North America and subsequently learnt from them by European settlers. Traditionally made maple syrup, we are told, requires on average forty gallons of sap to produce one gallon of syrup.

Maple 'sugaring', as the making of maple syrup is called, has been a way of life for Vermont farmer Burr Morse since he was a boy. Indeed, he is the seventh generation of his family to make it. 'With everything involved, it's very laborious. But we get it in our blood and we just love sugaring and we carry on doing it,' he laughs. 'I remember when I was five years old hanging around the sugar house, whittling sticks and mostly getting in the way. The sugar house atmosphere was just wonderful,' he tells me, with the affectionate warmth of that memory palpable in his voice. 'There was this sweet smell from the steam rising and sometimes we'd have doughnuts to go with the syrup.'

During his childhood, the maple trees were tapped and the sap collected using buckets. 'When I got to be old enough to carry two gathering pails, I'd be part of the gathering crew, who went from tree to tree, lifting sap buckets off the trees and carrying the sap from them to a sled with a tank on it. Some years, the first part of sugaring we'd be walking with snowshoes on top of three feet of snow. We'd tap as low on the trees as we could, kneeling right down because we didn't want the bucket to be out of reach when the snow melted.' This way of collecting the sap was as laborious as it sounds, he confirms, but there was camaraderie to it. 'That bucket system had fellowship to it. We needed a crew of four people to work together. We would joke and swear and laugh. Nowadays, with modern sugaring it's more solitary.' Morse fondly remembers one of the perks from the sugaring days of his youth. 'My father would be the boiler. Around two hours before the gathering crew came in at noon he would put hot dogs into a pot of sap over a flame, so the sap started to get concentrated and the sweetness would go right into the frankfurters and boy, they were good!'

Nowadays, in commercial sugaring operations, rather than using buckets, the sap is collected more efficiently, if less picturesquely, using miles and miles of plastic tubing – a system that came into general use during the 1980s – with vacuum pressure used to suck the sap from trees. The tubing has to be maintained, Morse points out. 'Before the season starts in the spring, you have to get into the woods and fix the tubing and use a chainsaw to cut trees that have fallen. It's a big job. We collect from around five thousand trees on around seventy-five acres of land.' The sap is deposited in tanks and must be transformed into syrup at once. 'You can't store sap,' says Burr adamantly. 'It's got just enough sugar content to start fermenting, so you want to make it into syrup just as quick as you can.'

The sugar content of sap, Burr tells me, varies from year to year and also within each season itself, starting off less sweet, rising to a peak in the middle of the season then falling off. On average the sugar content is 2%, though this year the sap has been very watery, containing only around 1–1.5% sugar. Historically, the sap was boiled for several hours to cook off the water and concentrate the sugar. Nowadays, however, many maple syrup producers, including Burr, use the process of reverse osmosis to treat the sap before boiling it in the evaporator. Reverse osmosis was developed to clean water; ingeniously, maple syrup producers use it to remove excess water from the sap, so increasing the sugar levels in the sap. An advantage of using reverse osmosis is that the fuel costs in cooking the sap down are reduced considerably. I'm told that whereas before, around four gallons of oil were required to produce one gallon of syrup, now, using reverse osmosis, it could take as little as one quart of oil to produce one gallon of syrup. Burr explains that while some other producers use reverse osmosis to concentrate sugar levels in the sap to 30% before heating it, on his farm they take the sugar levels up to only 12%. 'We feel that if you concentrate the sap too much through reverse osmosis, you sacrifice the flavour that comes through the oxidation of boiling.' Once the syrup reaches the right density, it is promptly stored in containers and sealed, whereupon it has a long shelf life.

Despite the use of new technology, one thing hasn't changed. Just as in the 'old days', the sugaring process is dependent on the right weather conditions being in place. The ideal temperatures, Burr explains, are around 25°F at night, ('no lower') rising into the 40s ('no higher') in the day. 'We can have a sap run and then lose our weather, which is the case today, and then we wait. We can wait for up to two weeks; I hope not but it's happened. When the sap weather returns, then we'll have another run. We've had seasons as short as one week and as long as six. It's absolutely up to the weather,' he tells me philosophically. The end of the sugaring season comes when the weather warms up sufficiently for the buds to swell, causing the sap to change chemically and making it no longer useful for making maple syrup. With the process of making syrup dependent on the weather, climate change is having an effect on maple syrup production. Research by the University of Vermont's Procter Maple Research Center has shown that the maple sugaring season on average now begins 8.3 days earlier and ends 11.6 days earlier than fifty years ago. Furthermore, the production and flow of maple sap is triggered by alternate freezing and thawing cycles, and a shorter season with fewer cycles within it mean that the trees produce less sap.

Grade A maple syrup – which must be clear, of a uniform colour and free from any off-taste – is classified into four categories: Golden Colour and Delicate Flavour, Amber Colour and Rich Taste, Dark Colour and Robust Taste, Very Dark Colour and Strong Taste. These categories reflect the stage in the sugaring season when the syrup was made, with Golden usually made early in the new season, Amber in the middle, Dark towards the end and Very Dark at the end of the season. The change in colour is due to the natural darkening of the sap as the season progresses. Intrigued, I invest in a set of three tiny chic bottles of Canadian maple syrup, consciously marketed as being syrups from different stages of the season, and sample them side-by-side. The first, a pale gold in colour, has a bright maple syrup flavour, followed by a pleasurable, mouth-watering acidity that reminds me of eating a piece of chocolate made from Criollo cacao. The second syrup has a fuller flavour, with no subsequent acid burst, while the last, labelled 'late harvest' is complex, with pronounced coffee notes and a longer finish.

Of course, though, the person best placed to enjoy maple syrup through all its stages is its maker. 'How would I describe the flavour of good maple syrup to someone who'd never tasted it? One word that comes to mind is heavenly,' says Burr simply. 'Imagine the best vanilla taste you've ever had and magnify it by ten. It's not overly sweet, it's a natural flavour that's hard to describe.' Each year as spring approaches he looks forward to that first taste of his own syrup. 'Maybe it's psychological, but for us the new syrup has a flavour that's hard to beat. And when it's just off the evaporator there's nothing like it.'

MONTHS

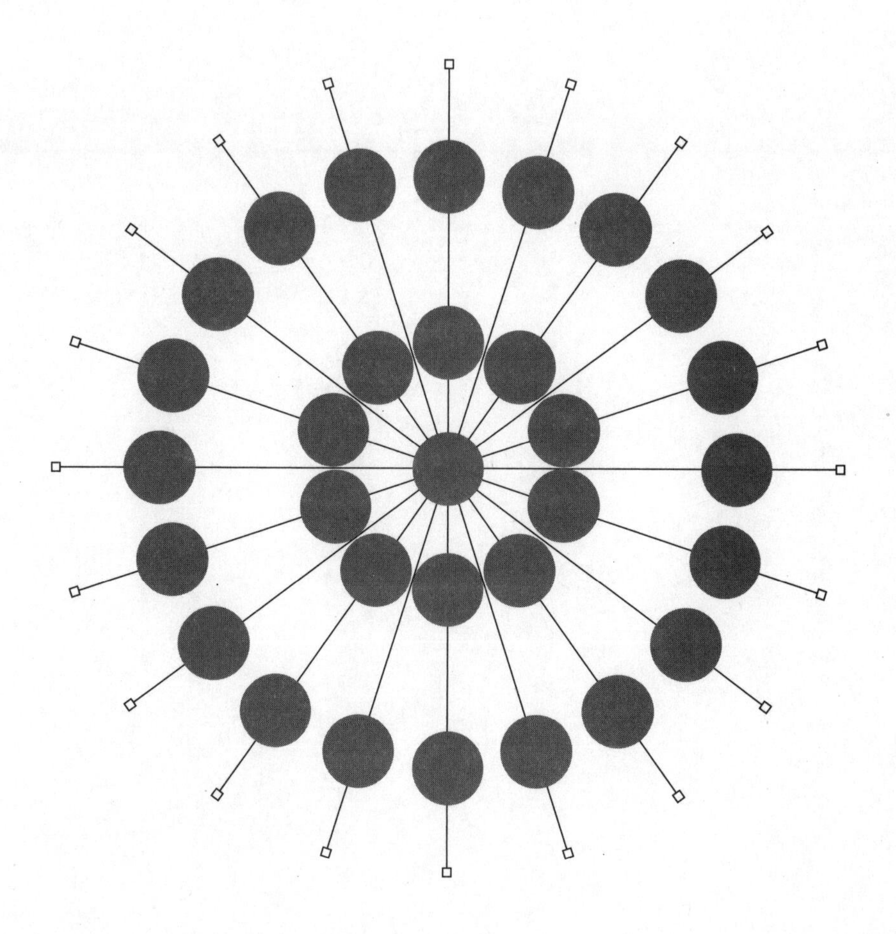

MONTHS

MONTHS

Weeks turn into months . . . We divide our calendar year into them and use them to pattern our lives. In temperate climates, these clusters of weeks present a way of marking the changing of the seasons: the passing from the short, dark days of December and January to the long, light days of June, July and August. Different months have historically offered seasonal foods – foods such as winter's citrus fruits or summer's luscious peaches and figs which were only available at certain times of the year. This idea of seasonality in food remains precious to us.

Preserving features here, from freezing's remarkable ability to keep food safely for months to the traditional use of fat as a preservative in dishes such as cassoulet and the ancient practice of steeping plants and spices to extract flavours and health-giving elements. There is a patience to those whose work spans months, an understanding that the time spent brings benefits.

A TASTE OF HONEY

Honey retains a special place in our affections. For thousands of years this natural, concentrated nectar was a highly prized source of sweetness. The world of honey archaeology, lovingly chronicled by the remarkable Eva Crane, shows how deep our desire for honey goes. A famous cave painting, thought to date from the Mesolithic period (approximately 10,000–5,000 years BC), shows human hunters collecting honey from a wild bee colony and is agreed to be the earliest depiction of humans and bees. There are written references to honey on Ancient Sumerian tablets and also in Ancient Egypt. It is a foodstuff that has a mystique about it – it is valued for health-giving properties when eaten and used to heal wounds. From a gastronomic point of view, honey ranges enormously in taste and colour, depending on the flowers that the bees have access to, so is valued for its expression of terroir. It is also noted for its keeping properties. A famous, much-repeated piece of honey mythology is that honey excavated from an Ancient Egyptian tomb was still edible centuries later.

Honey, of course, is made by honey bees as a source of food for themselves rather than for us. Such was its valuable allure, however, that with characteristic ingenuity humans set out thousands of years ago to keep bees, in effect 'farming' them, so as to easily harvest this unique and valuable sweetener that they produce. Whereas a natural bee nest, usually constructed in a hollow tree, is a striking, complex creation, beekeepers nowadays offer bees rather more prosaic hives constructed of straight-edged frames that allow humans access to the honey without killing the valuable bees in the process. Buying a jar of honey in a supermarket, choosing from an array of honeys sourced from around the world, it is all too easy *not* to appreciate quite what an extraordinary food it is. Handily packaged, ready to be simply scooped out using a spoon, the ease with which we human beings can consume honey belies the hard work that bees have put into creating it.

A visit in the month of June to Regent's Park, in the centre of London, brings home to me just how remarkable honey bees are. It is a fine summer's day and I walk along gracious avenues of trees, past carefully tended flower beds, making my way to a spot near the Inner Circle to meet Brian McCallum, beekeeper for Regents Park Honey. Toby Mason founded the company in 2005, part of the urban honey movement that has resulted in beehives being set up in central London – some situated on the rooftops of London landmarks such as Fortnum & Mason or the Tate Gallery – taking advantage of the city's parks and gardens as a resource for bees.

Practical and down to earth, as McCallum talks to me in his workshop his fascination with and respect for bees shines through. He describes with vividness and lucidity the intricate workings of 'the hive' or 'the nest', talking of it as a single living organism, made up of a cluster of 10,000 bees in winter, the population rising to 20,000 in the spring, then to around 40,000 in the summer. The rise and fall of numbers in the nest reflects the annual cycle of honey production. 'The nest does something that other insects don't do, which is that it lives through winter time. The reason that it can live through the winter is because it has stored the carbohydrate energy which is honey.' Plants, in need of pollination by winged insects such as bees in order to germinate, lure insects to them by producing nectar. The worker honey bees extract nectar from flowers, storing it in an internal sac where it is exposed to enzymes that will help transform it into honey, then carry it back to the nest. The honey created from the nectar is stored by the bees as a resource for the winter when it is too cold for them to fly and there is little nectar out there to collect. In the social structure of the nest, it is the worker bees who fly out to collect the food. 'They do a lot of flying, it's very hard work and then they die. Each worker bee has a mileage; it can fly about 500 miles and then it will die. They like to get their nectar nearby but they can fly anywhere between .6 miles and 1.8 miles.' In order to replace the bees that have died, there is 'constant regeneration', the queen bee laying up to 1,000–2,000 eggs a day in spring. In temperate countries, the annual cycle of honey production is triggered by increased amounts of daylight and a rise in temperature, with the worker bees first heading out after winter once it reaches around 54°F.

I am shown a frame from a beehive filled with a neat, patterned latticework of tiny hexagonal cells, made by the bees from wax they produce themselves. These cells are used as storage spaces, holding larvae, pollen or nectar. It takes time, McCallum tells me, to turn the thin, watery nectar into honey. The bees leave it for a month or two to evaporate, fanning their wings to aid this process, until the water content has been reduced to around 19%, at which point they cap the cell with wax. It is this honey, painstakingly created in the warm summer months, that McCallum will harvest in August, making sure, however, to leave enough honey in the hives for the colonies to live on through the winter.

Clad from head to toe in a beekeeping outfit, complete with a fine-meshed face mask and latex gloves – so that I resemble a deep-sea diver and feel as if I'm exploring a new world – I go with McCallum to visit the hives. They are tucked away in a peaceful, bosky corner of the park – surrounded by long grass and nettles – a gentle, pulsing, buzzing sound emanating from them. A steady stream of worker bees flies to and from each hive, some of the returning ones visibly laden with pollen. Carefully opening one of the hives, McCallum pulls out a frame to show me. It is a fascinating sight: the wax honeycomb inside is covered in a thick, living carpet of small, dark honey bees, crawling over it. 'They like to congregate together, to touch each other, know they're in contact with each other.' The bees remain calm in the face of our intrusion: 'They're gentle bees,' McCallum observes affectionately. They carry on their allotted tasks, unperturbed by being taken out of the hive. My overwhelming impression is one of busy industry, as in fables. 'They've all got something to do,' he agrees. McCallum points out the nectar stored in the cells, which the bees are waiting to dry out before they cap. An unusually cold spring has affected honey production, he observes, as in a good year there would be more. The main source of nectar within Regent's Park for these bees, he tells me, are the flowering trees; the lime trees, which are about to blossom, are especially important.

After I've said good-bye to McCallum I walk into the Inner Circle, passing through the elegant black and gold wrought-iron gates into Queen Mary's Garden in the heart of the park, home to the famous rose garden. The neatly planted beds are filled with flowering rose bushes in a palette of colours ranging from delicate, pale pinks to dark crimson. Peering at the rose beds more closely, I notice an absence. Not one honey bee in sight. I'm puzzled. Then I remember something McCallum told me. Flowers with overlapping petals, such as roses, are hard for pollinators to access. They prefer open-faced flowers, 'like the kind a kid would draw'. Around the edges of the rose garden

I notice hanging swags of climbing roses – white, open-faced roses. I stroll over to investigate. Sure enough, within a minute I've spotted a number of small, dark honey bees dancing round the blooms. I watch with a new interest as they nuzzle into the heart of each flower, moving deftly and restlessly from one rose to another. McCallum has told me that in summer a worker bee's lifespan is simply 20 days. As I watch them fly away, I wonder how many of their allotted days they have left to live. It pleases me, however, to think that Regent's Park, a green oasis for Londoners, is a haven for honey bees, especially at a period when bees are facing so many challenges from pests and disease.

McCallum gave me a small jar of Regents Park Honey, which I open when I get home. It is light golden in colour, possessing a distinctive scent. I look at it and marvel at the time and effort the bees have put into creating it. The flavour is far more intense than I expect from its delicate colour – not cloying at all, noticeably fragrant, with a long finish, lingering on the tongue. It tastes, I realize, like the perfume of linden blossom, my favourite scent – the taste of summer.

KEEPING FAT

Fat has historically been a foodstuff prized by humans, both preserved in its own right and valued for its useful keeping properties. There is archaeological evidence to suggest that we began eating butter, made from the butterfat present in milk, many millennia ago. Butter, however, is perishable, especially in hot climates. In India, clarifying butter, by cooking off its water and removing its milk solids, is a technique used to extend its life, mentioned in texts dating back 8,000 years. The resulting clarified fat, called ghee, has the advantage that it can be kept for months rather than days. As the cow is a sacred animal in Hindu religion, ghee is regarded as auspicious, used in temple offerings; 'the first and most essential of all foods', the *Vedas* called it. In Indian cuisine, ghee is used to add a touch of richness to dishes such as dhals, flatbreads and sweets. In other parts of the world it was a different cooking fat, lard, made by rendering down pig fat, that was once widely used. It remains a common cooking fat in places such as Mexico and Eastern Europe. So important was lard that in Hungary, the Mangalitsa pig, with its notably high fat-to-meat ratio, was reared primarily for the lard that it produced, rather than for its flesh. Animal fat, especially pork fat, plays an important part in charcuterie, used to add flavour and texture to products from fresh sausages to rillettes. In Italy's *lardo*, a charcuterie product made from the cured back fat of pigs, we see it valued in its own right. Tuscany's *lardo di colonnata* is particularly esteemed, made today as it has been for over 1,000 years: cured for several months with salt and herbs in marble containers in the village of Colonnata in the Apuan Alps. The resulting bright white *lardo*, usually eaten very finely sliced, combines a salty savouriness with a unique, luxurious melting texture.

Fat is also used for keeping food. In Britain the practice of potting foods, such as meat or cheese, under a layer of butter traces its origins to medieval pies. Correctly done, with the use of salt and spices, this allowed food to be kept safely for several weeks. Many foods, from grouse to lampreys, were formerly potted in this way, but today it is potted shrimps that remain as the best-known example of this culinary custom. On Morecambe Bay in Lancashire, the family-run company Baxter's, established in 1880, proudly carries on this tradition using locally caught tiny brown shrimps and a secret blend of spices in their melted butter. Among their devotees are the British royal family. In France, too, fat plays its part in creating delicacies, notably the dish known as *confit*, a speciality especially associated with the south-west of the country. The word derives from *confire*, which means 'to preserve'. With admirable practicality, it involves cooking cuts of meat and poultry from animals and birds naturally high in fat – pigs, ducks and geese – in their own fat. The tender, rich results are eaten either hot or cold and are often used in cassoulet, a dish made from haricot beans. Preliminary salting of the pieces is an essential part of the process, followed by long, slow, gentle cooking in fat at low temperature. The *confit* meat or poultry is then stored, classically in special earthenware pots called *toupins*, hermetically sealed with a layer of fat and kept in a cool, dark place.

A historic appreciation of fat as a valuable, useful ingredient diminished in recent decades, as fat became perceived as unhealthy. Butter and lard fell out of favour in Western kitchens, replaced by vegetable fats such as margarine and oils such as sunflower or olive. The unpopularity of fat has had implications for livestock rearing, expressed most clearly in the case of the pig, historically prized for its ability to put on weight. Today pigs in countries including Britain and America are bred to be lean. Changes in their diet, such as the move to soya, have also affected the quality of the fat that remains, lessening its keeping properties. Fat, however, is fighting back. There is fresh evidence regarding the healthy properties of previously demonized animal fats, such as butter. The rise of charcuterie producers in the UK is creating a market for animals that have the quantity and quality of fat needed in order to produce good salami or bacon. The demand from discerning independent butchers, and their customers, for flavourful meat from slow-grown native animals means that the presence of fat in meat is no longer taboo. Our recently discovered penchant for pork belly – a cut for which there are well over three million recipes on YouTube – is an expression of this. In her award-winning 2008 cookbook, *Fat*, the Australian chef and food writer Jennifer McLagan effectively champions what she terms a 'misunderstood ingredient', offering fat-centred recipes for cassoulet, dripping and terrines. The book's dedication is forthright: 'For all the Jack Sprats out there – you're wrong!'

STEEPING

It's a sunny day in May and I've returned from my walk with a carrier bag full of creamy-white, delicate, lacy elderflower heads, from which their distinctive muscat scent wafts up. I set to work, paring lemon zest, slicing lemons, making a syrup by boiling together sugar and water. The freshly made syrup is poured over the elderflower and lemon slices, wilting the flowers with its heat. Once cool, I cover it and set it aside overnight to steep. The next day, the syrup has taken on a pale golden honey hue and a fragrance from the flowers. I strain it and bottle the elderflower cordial to give to family and friends.

'Steeping' is a word which for me evokes medieval times, conjuring images of a monastery stillroom and a sense of calm and stillness. Steeping is something we do every day when making a pot of tea or cafetière – infusing the tea leaves or coffee grounds in water for a few minutes so that they impart flavour and character to the liquid. There is an ancient tradition, however, of more time-consuming steeping – taking days, weeks and months rather than minutes – in herbal medicine around the world.

Steeping is still central to herbalism, as I discover when I meet contemporary herbalist Michael Isted at his workplace in Lewes, Sussex. In the best apothecary tradition, on the table before him is a collection of Kilner jars containing concoctions filled with mysterious-looking leaves, flowers and roots. A dapper, youthful figure, Isted is, however, far removed from the stock image of a herbalist as an old man with a long, grey beard. 'When I meet people for the first time, they always say they were expecting someone a lot older,' he tells me, laughing. While working in the drinks industry, where one of the things he did was create flavoured syrups and spirits for cocktails, Isted became fascinated by the world of herbs, training first as a nutritionist, then as a herbalist. While inspired by the ancient texts of Chinese medicine, India's Ayurveda and the writings of the Persian scholar Avicenna, Isted is bringing his own way of doing things to his new profession. In his work he combines his experience, his knowledge and – importantly – his palate to create plant-based drinks, both non-alcoholic and alcoholic, containing powerful phytochemicals (chemicals found in plants which have protective or disease-preventive properties), which can be enjoyed in their own right. During my visit, I sample a number of his creations, from a delicate drink made from fermented green tea and honey to a spiced turmeric concoction, memorably complex and powerful.

As Isted talks me through the different steeping techniques, I realize just how complex and finely tuned the process is at this level. To begin with, he uses different materials to steep with – water, alcohol (at various strengths), glycerine, honey, vinegar, brine, oil. 'For me, it's about understanding the plant I'm working with, when to steep, what to steep it in and how long to steep it for.' Isted uses a range of techniques to extract different constituents from the plants he works with. 'This Roman camomile, for example, is in 95% alcohol, but there are constituents in it – such as charmazulene (an aromatic compound which has anti-inflammatory properties) – that will not leach out into alcohol, but will leach out into water.' Whether the plants are fresh or dried is another factor, fresh to Isted's mind giving by far the best flavour. Heat is used where appropriate to aid extraction, from sunlight used to create a calendula-infused oil to a water bath for 'tough, stubborn' turmeric roots, or adding hot water directly. On his mission to extract flavour and phytochemicals, Isted will apply a range of different steeping methods to one plant – using, for example, spring water and alcohol at different levels, from 20% to 95% – then combine the results. Working in this way allows him to get different flavour profiles and different constituents which he then blends together to create a 'picture and a story' of that plant. The tinctures he creates are used for medical purposes and also in drinks, to which they add powerful, concentrated flavours and aromatic compounds.

The amount of contact between the plant and the steeping medium is another crucial variable. 'The more time the better,' he says simply. There are minimum periods during which the steeping process does its work: at least a month for tinctures made using alcohol. Isted, however, enjoys experimenting where possible with longer steeping times. 'There's an old recipe for aloe vera steeped in rum with spices for 6 months. I made some 2 years ago and it just tastes better and better.' The Turmeric Plus that I sampled earlier has been steeping for 2 years, while the Roman camomile infusion is 3 years old. Working with hundreds of ingredients, as he does, means that steeping is something that Isted does every day, beginning the patient process of drawing out the properties from within the plants he carefully sources.

As a herbalist, working with the natural cycles, the season during which plants are picked is also vitally important. 'Spring is the prime time of year for me. I should be out gathering plants,' he says, giving a wistful look through the window at the bright May. 'This is the time when you've got those concentrated phytochemicals. Plants are bursting with vibrancy and energy for the new year and you really want to capture that.'

CELEBRATING SEASONALITY

Seasonality is a treasured concept in today's food world. A love of seasonality shines through much food writing across the board. Wonderful cookbooks such as Nigel Slater's *The Kitchen Diaries* and Pierre Koffmann's *Memories of Gascony* follow the pattern of a calendar year, evoking the changing pleasures as the months pass. In temperate countries, glossy food magazines, working months ahead, structure their articles around it, offering menus appropriate to the time of year, highlighting seasonal produce from winter's citrus fruits, such as Seville oranges, to autumnal game. Talk to the chef at any upmarket restaurant about the menu and chances are high that you will be told that it is 'seasonal'. High-end food shops such as pâtissiers and confectioners feature seasonal 'collections' of treats, inspired by spring, summer, autumn and winter. The deep human love of structure and pattern is satisfied at a profound level by the idea of seasonality.

Long before seasonality was a fashionable trope it was, of course, a simple fact of life in temperate countries. The availability of foods was linked to nature's seasons: new growth in spring, the abundance of summer and autumn, then lean times during the cold, dark winter months. This rooted reality meant that seasonality was not always fashionable; far from it. In the hierarchical medieval world, for example, the elite eschewed it. 'Living in a grand house, living in a royal palace,' says Marc Meltonville, food historian for the Historic Royal Palaces, when I visit him at Hampton Court, 'the whole point of the food here is NOT to be seasonal and local.' Courtly dining was about rank and status, and the food served to the court reflected this, using luxurious and exotic ingredients, such as spices brought from far away. 'The first thing that's unseasonal in a palace, from medieval times right up to the Georgians, is their access to fresh meat. That's the one we forget about. Fresh meat was seasonal for most people because they have only two killing months in the year: November and May.' The winter killing was of animals that had fattened over the summer and autumn, both to get meat for the winter and to reduce the numbers that would require feeding during the coming lean months, while the late spring culling was to thin out new animals and so make a stronger herd. In contrast, royal households ate meat throughout the year.

In the great houses much ingenuity was spent in growing produce that was unseasonal, for example creating sheltered hot beds. The kitchen gardens at Hampton Court had fireplaces along the walls, so that they warmed up the wall for the fruit trees on the other side, meaning they bore fruit a month early. An account of a Christmas dinner for Queen Caroline, George II's wife, during the 1730s gives an insight into the delicacies served, Meltonville says with a smile. 'When I read it out everyone gets very excited because there's a roasted turkey on the table. But everyone ignores the next line, which is about a large salad. It's the 25th of December and royal gardeners have managed to get a full mixed salad and asparagus onto the royal table for Christmas!'

Today, of course, lettuce at Christmas is no longer a novelty. A number of factors, such as refrigeration, global trade routes, airfreight and the use of polytunnels to extend growing seasons, mean that we have become accustomed to having everything all the time. Fresh fruits that were once only briefly available – such as strawberries – have become an everyday item, imported from faraway countries in warmer climes, when their availability here has ceased.

It is surely no coincidence that, just as the everyday experience of seasonality has blurred to become a rather monotonous, all-year-round abundance, high-end restaurants and chefs, seeking to offer something special and unique, have embraced foraging. Wild foods are, by their nature, truly seasonal, and this 'specialness' is a large part of their cachet. A hugely influential figure in this movement is René Redzepi of the world-famous restaurant Noma, founded in Copenhagen in 2003 with the specific mission to offer Nordic cuisine. Seeking a way of expressing a true, deep sense of place, Redzepi looked to the harsh landscape and seascape around him, working with expert foragers who hunt out wild foods such as moss, wild mushrooms, flowers, berries and spruce shoots. Many of these wild foods are painstakingly sourced, available only briefly, with their rarity, in effect, part of their status as a new luxury.

In Britain, chef-patron David Everitt-Matthias, of the renowned restaurant Le Champignon Sauvage in Cheltenham, has long used wild ingredients on his menus. He traces his fascination with foraged food to childhood visits to his aunt in

Suffolk when he was 7 years old. 'She was a good hedge-row cook,' Everitt-Matthias recalls fondly, 'and used to take me out with her to pick plants like elderflower, nettles, sorrel and damsons. It came back into my mind and I started going out myself around 25 years ago. We hired a herbologist to take us around the local lanes to show us plants we didn't know.' Everitt-Matthias talks with a real enthusiasm of foraged ingredients that he enjoys cooking with. Dainty scarlet elf cap mushrooms, which not only do not lose their distinctive colour when cooked but have a 'deep taste', he pairs with fish, using any broken ones in the fish sauce and sautéing whole ones as a garnish. Spring and summer sees the arrival of sea arrowgrass, with its 'wonderful taste of coriander', used to flavour ganaches and in fish dishes again. As he talks, he describes how he uses a plant throughout the seasons: early, tiny wild garlic shoots in salads, then, as the wild garlic progresses, blanching and mincing the long leaves to put through potato gnocchi or into a barley risotto. Wild garlic flowers are used in salads, then their seed pods are picked, salted for 48 hours, and pickled. 'They've got a massive punch of garlic, we keep them for winter, when we like to use pickles to lighten dishes.' Everitt-Matthias really does practice seasonal cuisine, rather than simply paying lip-service to the idea. 'When something's in season is when you should be using it,' he says with conviction. 'It's cheaper, it's plentiful, at its best and at its most flavoursome.' The pleasure of working with an ever-changing palette of ingredients is clear. 'It's like it's in, you use it and then it's gone,' he laughs.

An appreciation and understanding of seasonality is also at the heart of Natoora, a London-based British company that directly sources, imports and sells fresh produce. When I visited their warehouses, Franco Fubini, Natoora's founder, who spends much of his time travelling to growers, talked eloquently on how it underpinned his work. 'We have to follow the seasons, because what we do is source products based on flavour, so we try and find the best-tasting fruit and veg that we can. In order to do that, you have to accept the season and understand nature. When you eat a piece of a plant – it could be the fruit, the seed, the leaves, the root – these parts are at their best for our consumption within a certain time frame and that's really important.' Fubini sources direct from a network of growers in France, Italy and Spain. As he talks of his work, he paints a complex, detailed picture of the qualities of different fruits and vegetables and the skill and work required to grow good-quality produce. Rather than romanticizing all outdoor-grown produce, he pragmatically accepts the need to use polytunnels for certain crops, particularly tomatoes. Not only do polytunnels offer warmth and shelter to the crop grown inside them, but they also protect the grower. 'The grower needs the financial security that if he or she plants ten acres, a hailstorm won't ruin it. The benefit of the tunnel is that it can also protect from excessive rain.' He is unimpressed, however, with recent moves by British asparagus farmers to extend the season for their crop considerably by using polytunnels, feeling that the flavour is inferior. 'We draw the line at extending the season irrespective of taste. Tunnels should be used for the right reasons.'

With Natoora noted for the quality and flavour of its produce, I was surprised to see them championing so-called 'winter tomatoes'. They turned out to be Marinda tomatoes from Sicily, distinctively ridged, ranging in colour from green at the top to a deep orange body. Feeling sceptical, I sampled one – and was happily surprised. Tough-skinned and dense-fleshed, it had a noticeable depth of flavour, sweetness cut through with a lovely acidity in the way that the best tomatoes possess but one so rarely experiences. Fubini told me how innovations in growing techniques about fifty years ago brought about this particular crop. One of the factors that creates flavour in plants is stress. The winter tomatoes are grown slowly with very little water, something which growing in a cool season makes possible. 'You're trying to stress the tomato and take it to the edge of its life, so that it grows slowly and concentrates flavour. If you were to do that in the summer it would be challenging, because you would have to water it in order to keep it alive. The worst thing for this type of tomato is too much water. Also in the summer there is too much light and too much heat and they would grow too quickly.' Chefs and customers – like me, accustomed to thinking of tomatoes as being at their best in the summer months – were initially resistant, converted only through sampling. 'We coined the term "winter tomatoes" because the way it is grown means that it cannot be grown under that growing methodology in the summer. It is a seasonal product,' he says with satisfaction.

A brief season catches your attention in a special way. To begin with there is the charm of anticipation, of looking forward to something coming into season; next there is the satisfaction of enjoying the first that year; then, as it becomes plentiful, gorging on it; and, finally, there is realizing that the window to enjoy it has gone and you will simply have to wait another year. With the arrival of spring come two treats that have a special place in my affections. The first is Alphonso mangoes from India, noted for their smooth flesh, deep orange colour and noticeable perfume. This is the time of year when I go on a mango-buying expedition to Fruity Fresh's huge store on Ealing Road in Wembley, returning with a large box of Alphonsos to be devoured with relish by myself, my family and friends. Serving them simply involves cutting off the mango 'cheeks' and scooping out the soft flesh, which while intensely sweet is not cloying, cut through with notes of pine. Committed mango lovers then move on to suck the flesh from the skin and the large stone, relishing the particular flavour these parts of the fruit offer.

During the Indian and Pakistani mango season, Fruity Fresh has stacks of boxes of mangoes piled high, with customers coming in a steady stream to buy them, opening the lids and assessing them closely before making a purchase. While nowadays Fruity Fresh imports, wholesales and sells a huge range of fresh tropical produce, the business's roots and fame can be traced back to 1978, when its founder, Ashok Chowdry, began importing mangoes, then a rarely seen, little-known fruit. The trade grew in volume, creating the foundation for his successful exotic produce importing business. Over the decades, Ashok has seen the interest in Indian mangoes spread beyond the Indian community. Now the work at the Fruity Fresh warehouse begins each night at Western International Market, Southall, carrying on into the next day. 'At midnight the warehouse is buzzing, because you've got all the goods coming off the aeroplanes,' Ashok's son Neil Chowdry tells me. 'Once they come off the aeroplanes they are put on a lorry. When the lorry arrives at the warehouse, you've got to do stock control, check you've got what you should have, do quality control, separate out the products, send them out to the customers. I want to give my customers the freshest produce.'

During the mango season, around 3.31 short tons of mangoes will be processed every couple of days. Because Alphonsos are especially perishable, the emphasis is on processing them quickly, getting them to the shops within a few hours. Although Alphonso mangoes come in season in India in March, because of the high cost of airfreighting them Fruity Fresh waits until April, when the mangoes are more plentiful and the price per mango is lower before importing. Even then, they are an expensive treat. 'When we have a large volume of Alphonso mangoes, you smell it the other end of our market,' laughs Neil Chowdry. He muses on the special appeal of the Alphonso. 'When my wife and I have my in-laws round for dinner and we sit around with a box of good mangoes, it just brings everyone together. All the family are sitting round the table and enjoying the fruit; the little kids are going gaga because there aren't many fruits as sweet as a mango.' It is a pleasurable picture, which mirrors my own experience.

My other spring treat is, in contrast, a distinctly British one – namely outdoor-grown British asparagus, with its traditionally short season running from late April to 21 June. A visit in the month of May to Wykham Park Farm, Banbury, Oxfordshire, gives me an insight into the work taken to grow and harvest this luxurious vegetable. I have come at a very busy time of year for the farm, but fifth-generation farmer Lizzie Colegrave, an energetic, business-like young woman, kindly takes time out to show me round. It was her mother who first started growing asparagus on the farm as an experiment in 1991; today they grow fifty acres, with the bulk of it going to supermarkets. She drives me up the road to see one of their asparagus fields and we pass a group of pickers, on their way for a lunch break. The long field, with its orange-brown ironstone clay soil, stretches away from us. I am struck by how stark an asparagus field at this stage of growth looks. Rather than abundant leafiness, here are row upon row of short, green asparagus stalks, each spaced about a foot apart, poking up from the soil, looking somehow unexpected and slightly incongruous.

Outdoor-grown asparagus will begin to grow only when the temperature reaches 50°F, and the weather this year has not been good. 'It's really late this year, as it's been so cold,' Colegrave says, a note of disappointment in her voice. 'Asparagus is something that you can't force to grow. There is nothing we can do about it. It comes when it comes.' She is convinced that the flavour of British outdoor-grown asparagus comes from long, slow growing in a temperate climate. 'If it's a hot climate, it grows too quickly and I don't think you get the same intensity of flavour.' The reason for being disappointed that the asparagus has had a late start is that outdoor asparagus growers, like the Colegraves, have only a limited window in which to sell their crop, with the season's end dictated by nature rather than tradition. 'June 21st is the end of the season and the reason you stop picking then is that you want to leave about a third of the spears to grow, so that they grow up and becomes ferns,' Colegrave explains. 'When the ferns die down in winter, all the energy goes back into the crowns and that gives you the crop for next year's harvest. If you were to keep picking and picking until nothing came, you'd kill the crowns. In theory, these beds should last for 12 years.'

The asparagus spears are harvested only when they reach the right height, around 6 inches, with the picking knife used as a handy gauge. The fragile stalks, with their easily damaged tips, have to be picked by hand and picked every day, making it a labour-intensive crop to harvest and backbreaking work to boot. 'No one has developed a mechanized way of picking asparagus. Well, it wouldn't work, because you need to use your eyes.' She stoops and deftly cuts some stalks to show me: slender pale green spears, tapering to pale purple tips, truly a most elegant vegetable. In contrast to the long, patient wait for the first spears to come through, the growing season is a demanding period. One of the issues is that the rate of growth of the spears to a harvestable height is linked to the weather, with cool temperatures slowing it down and warm seeing it grow quickly. 'It's meant to be cold this weekend and that will stop it dead in its tracks really. And then you've got the issue of not having enough supply.' In contrast, hot weather brings an overabundance, with a pressure on the pickers to harvest it quickly enough.

Despite asparagus being a tricky, logistically demanding crop, the Colegraves are investing in further fields. I am shown a field planted the year before with 1-year-old male crowns, which will be able to be harvested in 2 years' time. The demand for asparagus has grown and grown over the years since that first one-acre trial crop. 'We put the signs up at the farm shop that asparagus is ready and it's like opening the floodgates. There's no other crop that does that. Asparagus is the only thing I can think of that people really go a bit mad for!' Colegrave laughs.

Before I leave the farm, I head to the large, professionally laid-out farm shop, on a personal mission to buy asparagus. A middle-aged woman enters the shop. 'Do you have asparagus today?' she asks anxiously. She is shown to the display, with its neatly arranged bundles of immaculate asparagus, and proceeds to put five or six of them in her basket. She catches my eye, as I'm watching her stock up. 'I love it,' she says, in explanation. 'Me too,' I reply. We smile in mutual understanding. Two shoppers enjoying a taste of seasonality.

THE FRESHNESS OF OLIVE OIL

For thousands of years, human beings have revered olive trees, first cultivating them in the Middle East around 6,000 years ago. In Greek mythology, the olive tree was a gift to mankind from the goddess Athena, its precious oil used to light lamps in temples, to anoint bodies and to cook with. With its distinctive flavours – ranging from sweet and fruity through grassy to pungent pepperiness – olive oil continues to be esteemed today, a cornerstone of Mediterranean cuisine. As with other oils, it has keeping properties, used in Italy for the *sott'olio* (literally meaning 'beneath oil') preserving tradition.

Traditionally, olive oil is a seasonal product, made annually by pressing the oil out of freshly harvested olives. The European season for gathering them usually runs from October through to January. The olives that are harvested earlier in the season, when they are less mature, yield less oil than those that are harvested later, when fully ripe. Early season oil, however, is generally considered to be a better, more flavourful product, with research showing that it has higher levels of polyphenols (plant compounds with antioxidant properties). The fact that olive oil is an agricultural product, varying each year according to the weather and growing conditions, gives an excitement to the first harvest. There is a sense of anticipation at an olive mill as the crop is delivered and cleaned before being pressed. The process of extracting the oil involves mashing the olives into a paste, pressing that paste, then using centrifugal force to separate the olive oil from the water and the solids. This new-season olive oil will then be sampled and analysed; the producer's hope being for a good yield of high-quality olive oil which can be certified as extra virgin (a grade reached through both chemical and taste testing). With its powerful aroma and often peppery kick, freshly pressed olive oil gives an extra dimension to simple, traditional dishes, such as Tuscany's *fettunta* (meaning 'soaked slice'), in which slices of rustic bread are grilled, then, while warm, rubbed with garlic, sprinkled with salt and liberally anointed with olive oil.

To enjoy extra virgin olive oil at its best, you would ideally sample it freshly pressed: 'It's much more vibrant,' says olive oil expert Judy Ridgway simply. 'If you taste an oil that's a year old but still decent, against a new-season oil, it will taste much flatter in comparison.' Once the olives have been harvested, let alone pressed, there will be a slight deterioration in the oil, though if care and attention is taken throughout the chain of harvest, processing, storage, transport and retail, this will be very slow. Extra virgin olive oils, explains Ridgway, deteriorate with time, eventually turning rancid.

The speed at which an extra virgin olive oil goes off, however, depends on a number of factors. Among these is the variety of the olives used, with olive oil made from Taggiasca olives deteriorating faster than oils made from Picual or Koroneiki. In assigning shelf-life, extra virgin olive oil is tested by the industry for its peroxide levels, with higher levels meaning it will degrade sooner. There is a cutoff point for peroxide levels past which the oil will not be sold. Ideally, therefore, one would want to buy an oil which had a very low peroxide level, rather than one with levels close to the cutoff point.

Processing, storage and display also have an impact on how the oil keeps. Frustratingly, for people looking for the freshest extra virgin olive oil, it is hard to find the relevant information. Very few extra virgin olive oil producers put the test results (including peroxide levels) or the harvest date on the label.

Instead, the date given is a use-by date, usually 18 months from bottling, but sometimes, observes Ridgway with disapproval, as long as 2 years. Since with larger olive oil packagers the oil may well have already been stored for months, that makes it even older than the use-by date suggests. Even if the oil doesn't actually become rancid, it will frequently deteriorate in flavour over the months. 'Often the fruitiness will fade, leaving you with much more bitterness and pepper,' Ridgway notes.

Encouragingly, Ridgway, who has been writing about olive oil for 25 years, feels that overall standards of extra virgin olive oil production have improved markedly in that period. When it comes to buying extra virgin olive oil, her advice is to choose dark bottles or tins, as clear glass speeds up the degradation process. Once a bottle has been opened, it should be used up as quickly as possible: 'Olive oil does not like air, heat or light,' she says firmly.

FREEZING

Throughout human history, people have sought to fight the decaying, destructive effects of time on food by preserving it in different ways. Among these techniques is freezing. This works as a preservative by transforming the water inside food into ice crystals, which in turn inhibits bacterial growth. In today's world, where freezers are an everyday household item, we take the ability to store frozen food simply and easily for granted. Much time and ingenuity, however, was spent in developing freezing technology. Drawing on their observation of the natural world, human beings have long understood freezing's potential as a preserving technique. In countries with very cold climates, such as Russia or the Arctic regions, freezing food by simply burying it in the snow was an everyday practice. For many centuries and in countries around the world, humankind's relationship with ice and freezing was a seasonal one, dependent on gathering and storing the ice that formed naturally in the winter months. Shady, cool, well-insulated structures were built to store the blocks of ice, allowing it to be kept for months without melting. In Persia, for example, distinctive structures known as *yakhchal* (ice pits) – subterranean chambers with domed tops – were built to store ice transported from the mountains in winter to the hot desert regions. In 1619 James I of England commissioned a brick-lined ice house to be constructed in Greenwich Park. Ice houses became a status symbol in England, built in the grounds of aristocratic great houses, allowing the elite to enjoy iced desserts during the summer months. The seventeenth-century poet Edmund Waller, writing in 1661, celebrated King Charles II's ice house in St James's Park:

Yonder the harvest of cold months laid up,
Gives a fresh coolness to the royal cup,
There ice, like crystal, firm and never lost,
Tempers hot July with December's frost;

The use of ice and snow were central to early experiments with freezing. In a well-known story, the seventeenth-century biographer John Aubrey in his *Brief Lives* recounts how the English philosopher, writer and scientist Sir Francis Bacon experimented with stuffing a chicken carcass with snow in order to preserve it, a portentous act given Bacon's death from pneumonia shortly thereafter. The reliance on naturally made ice continued for centuries. During the nineteenth century there was a booming trade in blocks of ice shipped from the USA to other countries around the world. Today at London's Canal Museum, visitors can peer into the cool, dark space of a brick-lined Victorian ice well, used to store ice shipped from Norway, a legacy of the building's construction in around 1862 as an ice cream warehouse for Carlo Gatti.

It was in the early twentieth century that an American inventor, Clarence Birdseye – his surname now recognizable from the corporation that brandishes it – made a significant breakthrough in the history of frozen food. Having lived as a fur trapper in Labrador, Canada, Birdseye had observed with interest the way in which the Inuit people used ice, wind and cold to rapidly freeze fish. This speedy freezing, he noted, resulted in minimal damage to the texture of the fish. Inspired by what he had seen, when he returned to America Birdseye set to work in 1922 to find a way of freezing food quickly, which he duly developed. The advantage of Birdseye's quick-freezing method meant that it produced small ice crystals, so that when the food thawed, its original cell structure was not significantly damaged in the process. 'I did not discover quick-freezing,' he wrote of his work. 'The Eskimos had used it for centuries and scientists in Europe had made experiments along the same lines I had.' What Birdseye did do, however, was make freezing food commercially viable.

The increased availability of domestic freezers, moving from expensive luxury in the 1930s to an everyday essential, has meant that access to frozen food is an everyday reality in many countries. So mundane, in fact, that frozen food manufacturers have to fight the perception that frozen food is lower quality than chilled. In fact, both freezing and chilling are ways of preserving foods that have their advantages and disadvantages. The rise of the microwave and, correspondingly, the growth of frozen ready-meals that can be quickly blitzed in them have, however, ensured the freezer's place in our home. You only have to look at the amount of space allocated to the freezer section in fridge-freezers nowadays – widely produced models offer 40/60 or 50/50 splits – to see our contemporary reliance on frozen food technology.

Freezing, of course, offers a truly useful way of being able to conserve many foods, though there are caveats. Bear in mind that many fresh fruits and vegetables benefit from cooking before freezing, as their texture and flavour are otherwise adversely affected; Birdseye's peas, for example, are blanched first. Blast-freezing technology, which freezes food very quickly, has benefits when used for fish or meat, as the rapid freezing prevents the build-up of dangerous bacteria. But as well as using freezers for fish and meat, foods that have a tendency to go rancid, such as nuts and whole-grain flours, can be 'stored' in the freezer, rather than in a cupboard. Being able to extend the lifespan of essentials such as milk, butter and bread by freezing them is a useful facility for domestic households. An essential aspect of running a freezer well is carefully labelling any package that you consign to its icy depths. I write this advice as someone whose freezer contains various mysterious frozen objects . . . foods that I thought, 'Oh, I'll remember what that is,' when I blithely placed it unlabelled, undated, in the freezer weeks, even months, earlier. These are the UFOs – 'unknown frozen objects' – found in many homes. A friend of mine cheerfully plays 'freezer roulette' before family holidays, adventurously defrosting a mysterious, anonymous frozen package for supper each night in the run-up, hoping to be pleasantly surprised.

Despite the ubiquity of freezer technology, ice cream – the quintessential frozen food – retains its long-held historical position as a special treat. Designed to be eaten while frozen, rather than thawed or cooked through like many other frozen foods, ice cream's alluring quality is heightened by its transience. The race to eat an ice cream cone on a hot summer's day, relishing the smooth cold texture before it melts into a sticky puddle, is a large part of the enjoyment of consuming it. Again, many years, much experimentation, skill and effort went into creating ice cream. In their fascinating book *Ices*, Caroline Liddell and Robin Weir emphasize the importance of the endothermic effect. This effect is created by adding salt to ice to depress its freezing point and transfer the cold by conduction from the ice to the mixture being frozen. A thirteenth-century account of the making of artificial snow using cold water and saltpetre (a chemical) by the Arab historian Ibn Abu Usaybi'a is cited by Liddell and Weir as the first technical description of making ice cream.

For centuries, however, because of the expense of ice and of sugar as an ingredient, ices and ice creams remained the preserve of a wealthy elite. It was during the latter part of the nineteenth century that ice creams and water ices become more widely available. Italian immigrants to Britain importantly brought with them their knowledge of how to make ice cream and began selling ice creams on city streets. In America, the development of an efficient ice-cream-making machine in the 1840s was a significant step in popularizing ice cream, and a decade later Jacob Fussell, a Baltimore dairyman, used his excess cream to manufacture ice cream to great commercial success.

Today, of course, commercially made ice cream is an affordable treat, mass-produced around the world. Flavours reflect national preferences, from Argentina's dulce de leche to Malaysia's durian and America's vanilla. Textures, too, range from the lightness of an Italian-style gelato to the luxurious creaminess of dairy-rich ice cream, in the classic American style. In London's Camden Market, a popular tourist destination, Chin-Chin Labs innovatively make and sell Europe's first 'nitro' ice cream, offering customers the drama of seeing their order made before their eyes, using clouds of liquid nitrogen. Made this way, the ice cream freezes so fast that the crystals formed are very small, giving it a noticeably smooth texture.

Kitty Travers has clear, firmly held views on the type of ices she wants to create. Having become fascinated by ice cream in her early twenties while living in the south of France, she spent years researching and experimenting in making gelato-style ice cream. In 2007 Travers set up La Grotta Ices, and since then acquired a devoted following for her distinctive frozen creations. Eschewing the rich, creamy, indulgent style, her ices and sorbets are at once sophisticated and enticing. Her range features intriguing, delicate flavours such as nectarine and lemon verbena, apricot and noyaux (apricot kernels), fig leaf and raspberry. The ices she produces are noticeably fine in texture, light and clean to eat.

A visit to Travers's small, self-styled 'ice cream shed' in Kennington where she creates her ices gave me an insight into her approach. The challenge she has set herself is to make ice creams with a particular texture, while using natural ingredients. Travers uses unrefined sugar, rather than glucose syrup, and natural milk and cream, rather than the milk powder, high in lactose, widely used by the ice cream trade. 'I want my ices to have the texture of Mr Whippy ice cream, that lovely, fake, amazing texture,' she says, 'but to be absolutely natural and fresh and fruity.' The seasonal use of fruit is central to Travers's repertoire. Having spent 7 years making ices and sorbets, she feels that the last 3 years were spent 'honing' her recipes. 'When something goes wrong,' she observes, 'it's always because you're trying to short-cut something.' As I watch her busily at work creating an experimental pink grapefruit, bay and white wine sorbet, Travers's pleasure at handling fruit is evident, from her descriptions of careful sourcing to the delighted smile with which she responds to the vivid scent of freshly zested grapefruit. Her production schedule is a weekly one, during which she produces small batches, the aim being to sell what she has produced. Her focus is on capturing fruit in gelato form, combining the flavour of fruits such as Gariguette strawberries, mulberries or papayas with the seductive texture of ice cream. Her inspiration is deep-rooted: 'I always think of when I was little, reading *James and the Giant Peach* and the bit when he crawls out and starts eating mouthfuls of this pure, lovely peach. That's always in my head, when I make my ices.'

Each batch of ice cream is made by Travers with great care and attention to detail throughout the process. Her quest for quality of flavour sees her using time in different ways to create what she wants. Syrups, for example, are made with an awareness of ingredients and their impact. I watch as, having simmered a fragrant bay leaf syrup – bearing in mind, she tells me, that the flavour of the bay will be muted by around 15% by the freezing process – she then adds grapefruit zest, but allows the zest only a brief contact time with the syrup, as otherwise the grapefruit's natural bitterness would be too dominant. Although the bay leaf syrup was made through heating, many of her syrups are made by cold-infusing for several days. Elderflower, for example, she cold-steeps, as it allows for a better flavour than hot syrup, which, she says expressively, 'gives that boiled-cat-pee taste to it'. Time, it becomes apparent, is also used in preparing both the fruit and the custards used as a base for her ices. The fruit is macerated with sugar and an acid such as lemon juice and left overnight in the fridge. Similarly, the custards, made by hand from egg yolks, milk, cream and unrefined sugar, are 'matured' in the refrigerator for at least 8 hours, a process known as 'ageing the mix', which she recommends to anyone making ice cream. Ageing the custards in this way 'dramatically changes the texture of an ice cream, gives it more body, means it scoops better and holds its shape in the cone'. When teaching her popular ice-cream-making course at the respected School of Artisan Food, Welbeck, Nottinghamshire, Travers gets her students to make ice cream using both freshly made custard and matured custard so that they can see for themselves the difference it makes. The macerated fruit is then whizzed, strained, combined with the aged custard, whizzed again, then frozen. At this stage, because there is no air in it, these blocks of ice cream mixture can be kept for several weeks without loss of flavour.

The final stage of the process is the churning, for which Travers uses a Pacojet, widely used in restaurants. A sophisticated piece of equipment – in contrast to her stainless steel pail, chinois and ladle – this, in effect, finely shaves tiny pieces of the frozen mixture, creating a smooth, soft texture. Each litre of ice cream is churned in 4 minutes, then packaged up for sale. Taking a solid frozen mass of ice cream mixture from her freezer, Travers demonstrates how it works. With a roar the Pacojet transforms the top layers to a portion of soft ice cream. I am experiencing a privileged ice cream moment, eating ice cream seconds after it has been freshly churned – 'The best way to eat it,' smiles Travers. Happily for me, we sample an array of flavours, from a refreshing strawberry salad creation, made from strawberries, blood oranges and lemon, to a richer-textured buffalo ricotta with pistachios and candied peel, a riff on Sicily's cannoli. While the colours are pastel, the flavours are clear; even in those ices where Travers has combined flavours, each is present and discernible. I am struck afresh by freezing's ability to capture flavour and suspend it in time.

THE BEST BEEF

If asked to name some traditional British dishes, the chances are high that roast beef would be on that list. It is the meat of Sunday lunchtimes: a splendid piece of roast beef – cooked so it is still pink inside – served with roast potatoes, Yorkshire puddings, gravy and piquant horseradish sauce. Beef and Britishness have become entwined over the centuries. Britain's climate and lush pastures make it a country well suited to the rearing of cattle, enabling Britons to develop and sustain a taste for beef. In animal husbandry, feed conversion rates are used to measure how efficiently livestock convert their feed into the desired output. With a feed conversion rate of 8:1, as opposed to 3.5:1 for pork, beef has quite properly long been seen as a luxury. Prime cuts – representing just a fraction of an animal's whole carcass – are particularly cherished: rib of beef, fillet, sirloin. So valued is the meat that we have developed a way of maturing it, hanging the carcass in cool conditions so as to enhance its flavour. Talk to butchers and chefs about what makes good beef, however, and the conversation will quickly turn to the live animal and how it has been reared.

The quest to learn more about how beef is reared took me up to Cheshire. Waiting, as promised, to meet me at Crewe railway station was the tall figure of farmer Keith Siddorn. During the 40-minute drive to his farm, as Siddorn talks about his life, farming and the fertile quality of the soil in Cheshire, I realize just how rooted he is in this particular part of the world. He comes from farming stock, the fourth generation of his family at Meadow Bank, a 200-acre mixed farm. 'If you look in the telephone directory, there's about thirteen Siddorns listed and we're all farmers and we're all within five miles of here.'

Once at the farm, after a restorative cup of tea, we head out to meet Siddorn's herd. On this bright, sunny October day, the sight of cows grazing peacefully – ruminatively – on lush green grass in a field is bucolic, befitting the pastoral surroundings. What seems to be an everyday sight is, in fact, remarkably unusual. These particular cattle – small in size, possessing dark red coats and distinctive white markings – are rare Traditional Herefords, one of Britain's oldest native breeds, originating in Herefordshire in the mid-1700s. They are now endangered. There are only about 700 Traditional Hereford (also called Original Population Hereford) breeding females in the world, of which Siddorn's herd forms 1%. Each animal here can be traced back through the herd books to 1846, when the herd book defining Hereford was first set out. Historically the Hereford's capacity to thrive on grassland made it a very popular breed of cattle, exported from Britain to countries around the world, including Australia, North America, South America and South Africa during the nineteenth and twentieth centuries. As the cattle travelled, it was cross-bred, leaving merely a small gene pool of the original intact.

There were a number of reasons why Siddorn chose Traditional Herefords when he set up the herd on the family farm in 2001. He was interested in genetics and breeding and liked the idea of doing something to preserve an endangered breed. Since he started the herd, there have been 507 Traditional

Herefords born at Meadow Bank. Through his breeding work, carried out with the Rare Breeds Survival Trust, he seeks to improve the cattle's traits, such as maternal instinct, hardiness and (for the beef) more internal fat and marbling, in order to make the breed more commercially viable. His respect for Traditional Herefords becomes apparent as he talks about their qualities. Having initially had a herd of Limousin and Charolais – modern breeds that don't like living outside and require being fed corn to supplement their diet – he looked for something hardier. 'We wanted cattle that live outside 365 days of the year. It's not natural for an animal to be in a shed all of its life, fed a high-protein diet. In winter, because the land's too wet, we put them on arable fields and give them the feed we've conserved during the summer: silage and hay. April time they go back on the grass.' Whereas modern-day cattle fed indoors in big feed lots can be finished in 16 months, Siddorn's Traditional Herefords put on weight more slowly, taking around 24–30 months to finish. While a typical Limousin carcass at 16 months would be 726lb dead weight, his cattle would be 517lb: 'So a hundred kilos [220lb] less and the fat content – the marbling in the meat – far higher.' This latter, in Siddorn's view, is an advantage. Herefords, unlike dairy cattle, were bred specifically to produce meat. Modern breeds, lacking fat, offer lean meat, which is 'tough, dry and doesn't taste of anything'.

Another desirable attribute of the Traditional Hereford, from a husbandry point of view, is its even temperament and its good maternal instincts. As just Keith and his father, John Siddorn, look after the farm, they needed cows that were safe to work with, not 'nowty' (a Cheshire word meaning 'naughty and nervy'). Walking through the pasture with Siddorn, I see for myself just how calm they are. Despite the fact that we are passing very close to mothers with calves – and that Siddorn's young dog is overly excited and keen to chase them – the herd gaze at us peacefully, unperturbed by the intrusion.

The welfare of his herd is important to Siddorn, central to the way in which he farms. He waits until the female cows are 3 years old before they give birth, in order to make sure that they are fully matured. The calves are left with their mother for 8 or 9 months. Rather than routinely cull cows at 8 or 9 years, if they are still breeding they stay in the herd until they die naturally or sickness means that they have to be put down. 'She's an amazing cow,' he says affectionately, pointing out a particular animal to me. 'She's 17 years old. Very productive and a good mother; we've never assisted in a calving. She will die on this farm and then I'll have to pay £150 to the knacker man to take her carcass away.'

It is not just the cattle on his farm that he looks after. Siddorn's interest in the environment has resulted in him planting hundreds of trees, miles of hedgerows and digging out ponds; undoing, in fact, much of what was done by his father in the 1970s when he moved from dairy farming to arable. His focus on animal welfare and the environment are unusual; 'A lot of other farmers think I'm mad,' Siddorn tells me matter-of-factly. He points out, though, that of the twelve local farming friends he went to school with, he is the only one still making a living from farming.

As with any farmer, though, there is a lack of sentimentality to Siddorn. On our tour of the farm, he remarks, 'I'm proud of my cattle. It makes my mouth water to look at them.' One of Meadow Bank's income streams comes from his Hereford beef, produced from the males. These are slaughtered at 2 years, since BSE legislation stipulates that cattle over 30 months must not be sold on the bone, which means that the most profitable cuts – rib of beef and T-bones – cannot be sold from older animals. As we walk among the herd, Siddorn points out the shape that he is looking to achieve through breeding. It is the back end of the cow that is important for meat production: 'All the value is from the ribcage back. So we want animals that have a good rounded back end.' In order to survive, Siddorn points out, a rare breed needs to be commercially viable. His cattle are slaughtered and skilfully butchered by Callum Edge, a sixth-generation butcher who runs a butcher's shop and abattoir in Birkenhead. Working with Edge is important. 'All my hard work – choosing the right breed, those years of breeding the cow, waiting for it to calve, waiting for the calf to come on line – it could be ruined in the last few weeks. Poorly slaughtered, not hung correctly, not butchered correctly – you can ruin a good piece of meat very quickly.'

An example of good butchering is that Edge cuts the hide off, rather than pulling it off, allowing the subcutaneous layer of fat to be retained on the meat. The neatly butchered beef is returned to Siddorn's farm, since he is the sole retailer of his meat. Rather than being showcased in an upmarket butcher's, Siddorn sells his meat from a small portable cabin just by the farmhouse. The plastic-wrapped cuts – from mince to ribs – are stored chilled or frozen. Over the years, without resorting to a website or social media, he has built a loyal customer base for his beef. Despite the shop's modest appearance, Siddorn's firm convictions regarding the way he farms and the quality of what he is producing are made clear on a sign in the shop: 'By purchasing this piece of succulent beef you will enjoy the most tender and deliciously flavoured beef you have ever tasted as well as helping us to conserve this most rare and important breed of cows.'

YEARS

YEARS

YEARS

Very few foods can be matured successfully over years rather than months. Those that can – among them Spain's jamón ibérico and Italy's Parmigiano-Reggiano cheese – enjoy a special cachet as delicacies to be sought out and savoured. These are foods that command high prices for the skill and considerable investment of time that has gone into creating them. Strikingly, alcohol, with its special capacity to age well, comes into its own. The unique mellowing capacity of time – seen in the lengthy maturing process that transforms a harsh spirit into a mellow, golden-coloured whisky or a youthful fortified wine into a heady, fruity port – cannot be mimicked. The idea of age is bound up in the way in which aged spirits and vintage wines are presented. The knowledge that so many years have passed in creating drinks such as a 30-year-old bottle of Pedro Ximénez is one we instinctively respect and find interesting.

PERFECTING PARMIGIANO-REGGIANO

A whole Parmigiano-Reggiano cheese is an impressive sight. There is a satisfying solidity to the large drum-shaped wheel, which weighs at least 66lb. The tough natural rind is golden in colour, stamped with information – the name of the cheese, the date of manufacture, a quality stamp – that authenticates it. 'No other Italian cheese is as large as a *forma di parmigiano* – a whole Parmesan – and no other cheese in the world is aged for so long,' observes Italian food writer Anna Del Conte in her authoritative *Gastronomy of Italy*. One of the world's great cheeses, Parmigiano-Reggiano has been produced in the verdant Parma-Reggio region of Italy for at least 8 centuries. In 1348, in his *Decameron*, the great Italian writer Boccaccio playfully described the fantastical land of Bengodi, a place of plenty. Here sausages are used to tie up vines, a goose and a gosling can be bought for a farthing, and people live on a mountain of grated Parmigiano. It is a cheese to which the Italians have long had a special attachment, bringing it with them on their diaspora journeys to countries such as America.

The making of Parmigiano-Reggiano nowadays is a closely regulated affair. Strict rules regarding all aspects of its production have been laid down by the Consorzio del Formaggio Parmigiano-Reggiano, formed by cheese-makers in 1954 to protect their traditional processes. These stipulate the area where the cheese can be made, the feed given to the cows, the use of raw milk, how it is made and the ageing period for different grades. After 12 months' maturing, the cheeses are inspected and the rind branded according to the category they fall into. The youngest is Parmigiano-Reggiano Mezzano, eaten soon after the minimum 12 months of ageing. After this stage there are three recognized ages: over 18 months, over 22 months and over 30 months.

Each cheese requires around 145 gallons of milk to make one wheel. This majestic size enables the notably lengthy maturing time. The *minimum* age for Parmigiano-Reggiano is a year. It is made from a mixture of skimmed milk from the previous evening's milking (left overnight, during which time the cream rises to the surface and is skimmed off the following morning), mixed with whole milk from the morning's milking. The overall low fat content contributes to its distinctive dry, granular texture. Parmigano-Reggiano and the idea of age are closely bound together. Innovative Italian chef Massimo Bottura, of the world-famous restaurant Osteria Francescana, is known for his playful culinary explorations of the food heritage of his native Emilia-Romagna. One of his celebrated dishes features Parmigiano-Reggiano, matured for different periods of time and presented in various forms: a demi-soufflé, a galette, a sauce, 'air' and a foam. He calls it the Five Ages of Parmigiano-Reggiano.

Today there are around 350 dairies producing Parmigiano-Reggiano cheese, ranging from small, artisan operations to larger ones. Given the level of regulation, what is striking is that rather than uniformity there is a noticeable variety of flavours and textures on offer. John Elliott and Alison Crouch, whose business, the Ham and Cheese Company, imports Italian cheeses and charcuterie into Britain, can testify to this, having spent 2 years visiting at least seventy dairies to find the cheese they wanted. 'We didn't even know if what we were looking for existed,' muses Elliott, when I visit his south London warehouse. The Parmigiano-Reggiano they have now been importing for 7 years is made on a very small scale by a producer using milk from his own cows: 'Umberto makes just three wheels a day. This way he can control the cheese-making, even when the quality of the milk varies. Some of the larger dairies make 150 wheels a day.'

Elliott is wary of the conventional notion that the older a Parmigiano-Reggiano is the better it will taste. 'There's a shop in Bologna that sells 10-year-old cheese. It's like a museum really. Who would want to buy 10-year-old cheese?' The cheese he sells is aged for 24 months: 'We wanted a Parmesan that could be used as a table cheese, as well as for cooking, as they do in Emilia.' He points out that nowadays Parmigiano-Reggiano is predominantly made using milk from Friesian herds, rather than the traditional breeds used several decades ago. In Elliott's opinion this fact has a direct bearing on successful ageing: 'I think it's a mistake to take the milk from a Friesian and age it for a long time, based on an idea from a different period in history that older is better.' I have brought with me a selection of Parmigiano-Reggiano cheeses bought in shops ranging from a supermarket to a cheesemonger, so that we can do a comparative tasting. The results vary considerably in aroma, texture and flavour. 'This one has that characteristic smell, that vomity smell,' observes Elliott matter-of-factly. Finally, I try a piece of the Parmigiano-Reggiano imported by the Ham and Cheese Company. The scent is pleasantly milky, a gentle aroma, very different from the other cheeses. The flavour is sweet to begin with, but developing a savoury saltiness with a noticeably long finish. White dots of cheese crystals, formed in the cheese as it has aged, give a crunch, contrasting with the soft paste. 'This is the Parmesan we were looking for,' says Elliott with satisfaction.

The cheese is an umami-laden tribute to the quality of the

milk, the skill of the cheese-maker and the lengthy maturing process. It is this depth of flavour, created over time, that makes Parmigiano-Reggiano so prized. Best as a grating cheese, it occupies a central place in Italian cuisine, but 'always used with discretion', observes Anna Del Conte sternly. It is worth noting that cheese grated freshly from a chunk, rather than packaged pre-grated cheese, has far more flavour. The sprinkling of a little freshly grated Parmesan cheese over a portion of *tagliatelle al ragù* or stirred into a risotto enhances and enriches. 'The whole idea of Parmesan is that it lifts a dish,' Elliott remarks.

WHISKY HERITAGE

Age and ageing are important in the making – and marketing – of single malt Scotch whisky. The spirit's name comes from the Gaelic *uisce beatha*, meaning 'water of life'. The first written record was an order for it from King James IV in 1495. The taste for whisky has spread around the world and Scotch whisky now plays an important part in both Scotland's and the UK's economies. Single malts, as opposed to blended whiskies, rose to prominence outside Scotland during the 1980s. Maintaining standards in this sought-after spirit has been important. In 2009 legislation was brought in that defined five categories of Scotch whisky: Single Malt Scotch Whisky, Single Grain Scotch Whisky, Blended Scotch Whisky, Blended Malt Scotch Whisky, Blended Grain Scotch Whisky. Legally, Scotch whisky must be matured in oak in Scotland for a minimum of 3 years. Maturation in wooden casks means it becomes less alcoholic, as it loses what is termed the 'angel's share' through evaporation. As the spirit ages in oak it takes on colour and flavour from the wood, giving whisky the golden-brown hues associated with it. Spirit caramel (E150a) is a legally permitted additive, though, allowing producers who wish to do so to replicate the colour that long ageing would naturally give the spirit. The age given on a bottle has to be the minimum age of the whisky inside, so a 12-year-old whisky contains whisky within it that is *at least* 12 years old.

In today's buoyant whisky market, extra-aged whiskies, i.e. aged in cask for 25 years and longer, have acquired a special cachet. Rarer and more expensive by their very nature, old whiskies are collectable and sought after. A look at one of the numerous whisky auction websites reveals that they command prices in the thousands; an eye-watering £62,600 being the successful bid for a 50-year-old Yamazaki, produced by Japan's oldest whisky distillery. At a less exalted level, age statements – 10-year-old, 12-year-old – are a customary sight, but Sukhinder Singh, co-founder and owner of whisky retailer the Whisky Exchange, cautions against automatically assuming that the older a whisky is the better it is. Singh observes that many of the whiskies highly rated by knowledgeable experts are young, between 10 and 18 years old. 'People need to use their own judgement when it comes to whisky,' he says adamantly. 'Older whiskies have spent more time in the cask, so the wood influence is greater. If people like gentler, cleaner, fresher notes, then they won't like older whiskies.' The pressure on distilleries to produce enough to satisfy a voracious, growing market has seen the rise of No Age Statement (NAS) whiskies, whiskies which don't bear an age on the bottle. Macallan, a Scottish distillery famed for its single malts, is among the whisky producers offering such a range.

The story of how Scottish distillery Bruichladdich makes whisky offers an insight into the role that time plays in single malt production. The distillery's own story reflects the roller-coaster ride Scottish whisky distilleries have been on. Established on the island of Islay in 1881, the distillery was sold during a period of acquisitions and in 1994 it was closed by its then owner. Bought by whisky enthusiasts in 2000 and reopened, the distillery is nowadays noted for its open-minded, innovative approach to whisky-making. 'We're less interested in the idea that older means better,' declares head distiller Adam Hannett, himself an Islay man. Despite that, the care taken in making whisky at Bruichladdich is manifest. 'We distil very slowly because that way you get a good texture and preserve the oils.' The advantage of slow distilling is ample contact with the copper still, since, if not 'pushed too hard', the vapour condenses at the cooler neck of the still and runs back down, creating an oily spirit that carries flavour well. This first distillation is then distilled for a second time in the spirit still, a process during which timing is very important. The first stage of this second distillation produces very alcoholic, pungent 'foreshots'; the middle stage is called the 'heart', and the final stage the 'feints'. Only the alcohol produced in the middle stage is used by distillers to make whisky. Each distillery has a different time for the foreshot. At Bruichladdich they run the process first for 35 minutes before extracting the middle cut. 'Ours is 35 minutes before we extract the heart because we want to get rid of any volatiles that won't give us good flavour.' When it comes to the crucial stage of maturing the whisky in wooden casks, Hannett is experimenting. While bourbon casks are traditionally used in the whisky industry, Bruichladdich have just received a delivery of Marsala casks. 'We've never used these before, so it's exciting for us to fill them and see what happens. Yes, it's a gamble,' Hannett acknowledges cheerfully. For Hannett, careful distilling, focused on quality of spirit, gives Bruichladdich the possibility of creating good whiskies sooner than conventionally accepted. There is, it seems, a youthful energy in the world of whisky, with ageing no longer synonymous with quality.

RETURN OF THE NATIVE

For anyone interested in biodiversity, a look at the Rare Breeds Survival Trust website makes for sobering reading. In the twentieth century, between 1900 and 1973, when the trust was formed, the United Kingdom lost twenty-six out of 140 of its native breeds. As farming moved to a more intensive, industrialized model, farmers shifted to modern breeds (also known as continental breeds), bred to put on weight fast, and slow-growing native breeds fell out of favour. It was the stark realization that many native breeds, part of our cultural heritage, were endangered that prompted the Rare Breeds Survival Trust to be established, in a bid to conserve them. Cheese-maker Charles Martell, for example, recalls the situation with regard to Gloucester cattle, from whose milk both Single and Double Gloucester cheese was made. 'In 1972 there were only sixty-eight left in the world – a puff of wind and they'd have gone. Now there are around 700 females, which still isn't a lot,' he observes. While the situation has improved for certain breeds, the Rare Breeds Survival Trust watchlist reveals that the situation is still precarious for many.

Talking to good butchers and chefs who work daily with meat, the consensus is that traditional native breeds offer the best eating. This is not mere sentimentality or nostalgia, but rooted in biological realities. A vivid, forthright speaker,

Cumbrian-born butcher Andrew Sharp, a founding member of the Mutton Renaissance campaign, talks me through some of the differences between historic and more recent livestock breeds. A traditional beef shorthorn would take at least 4 years to reach 280kg for a carcass. In contrast, a modern Belgium Blue ('the Arnie Schwarzenegger of animals') would reach that weight in 18 months. Similarly, with sheep, while a Herdwick (one of the 'primitive' breeds) would take around 18 months before it was finished for market, a Texel would take just 6 months. 'The difference is speed,' says Sharp plainly, 'but it's like anything on the farm – animals or vegetables – if you produce it fast, it will have high water content and a low flavour profile because it is dilute.' Slower-growing livestock, of course, involves an extended period of expense for the farmer rearing it, from worming costs to buying feed. On the plus side, however, the native breeds are hardier, able to feed successfully in harsh landscapes like fells, whereas modern breeds require a lot more input, such as inside housing and cereal feeding. Fat is one of the major factors in creating flavour in meat, but it needs to be fat that has been, in Sharp's phrase, 'naturally earned'. The fat created in cattle fed on corn will have less flavour and texture to it than grass-fed fat, which will also be higher in omega-3. 'The right fat, gained in the right way, is a far different thing from an animal just being fat.'

Sharp is encouraged by the rise in interest among farmers and butchers in native and rare breeds, as meat-shoppers, prepared to pay for quality, begin to understand and value them. He contrasts the eating experience offered by a Texel lamb and a Herdwick lamb. The first, ready for slaughter at 6 months, is 'very tender, but watery, and if you blind tasted, some people might think it was tofu with a bone tied to it'. In contrast, a

Herdwick lamb would only be ready to be eaten an additional 12 months later. The meat, says Sharp with relish, 'will eat denser, the fat won't have any tallow flavour, it could be cooked rare and it would have a bigger, savoury taste'.

THE CREATION OF SHERRY

Time is used in an intriguing way in the creation of sherry, a Spanish fortified wine still made using the *solera* system, developed in the eighteenth century. This cask-based system of production is, in effect, a constant cycle, in which the wine moves through a series of wooden casks as it matures. The easiest way to understand it is to picture four tiers of casks on top of each other. The process involves the wine being moved through the tiers from top to bottom. When the wine in the bottom tier is ready to be bottled, a portion of it is carefully drawn off, and replaced with wine that is slightly less old, from the tier above. This turn is replenished from the tier above that, and so on. The final tier is topped with new wine, called *sobretabla*. 'Fractional blending' is the technical term for this way of making wine. The image of the stacked casks offers a way of understanding what is a complex process. The reality is that larger *soleras* operate on a big scale, with entire blocks of casks being treated, in effect, as one tier. Regulations today stipulate that at the most 35% can be taken out at any one time, but usually a smaller percentage is withdrawn. The *solera* system means that the resulting sherries contain wines from a range of periods, skilfully blended together to create specific profiles for the various sherry styles.

The different sherry types are created during the maturing period, with two different kinds of ageing used. The use of a yeast layer called the *flor* is an important part of sherry-making, used in 'biological ageing' (the term used for sherry developed under this layer of *flor*). In order to produce *flor*, the huge casks are deliberately not filled to the brim, leaving space for oxygen in order to allow the yeast, naturally present in the atmosphere, to grow. Furthermore, the alcohol levels in the fortified wine are set at a level which permits the *flor* to develop. In ideal conditions, the yeast inside a barrel would grow and cover the surface of the wine completely in 3 weeks. This 'blanket' of yeast prevents any oxidization taking place, resulting in noticeably pale wines such as Fino, which does not darken even after it has been aged for 4 years in cask. As the yeast dies, it falls to the bottom of the cask and breaks down, releasing nutrients to feed the next generation of yeast, so creating a cycle of renewal. The resulting Fino sherry is delicate in flavour and austerely dry, classically enjoyed chilled as an accompaniment to savoury *jamón* or olives.

The other type of ageing is oxidative. This also takes place in wooden barrels, but without the creation of *flor*, as the wine is fortified with spirit to a higher level of alcohol to prevent yeast growing. As the wine ages in wood, it oxidizes, darkening in time. This process is used to create Oloroso sherry, dark in colour, with a nuttiness to its flavour.

The sherry houses work with these methods to produce a range of wines, including Amontillado, which is created using both biological and oxidative ageing. Given the *solera* system, the sherry producers generally do not label their wines with the year of harvest; only rarely will one find a vintage sherry. When an age for sherry is given, it is based on a calculation involving the number of scales (groups of barrels) in a *solera*, how often the wine is taken out, and how much.

Visiting Jerez de la Frontera, the town in southern Spain after which the wine is named and a notable part of the Sherry Triangle where it is made, I grasp the importance of sherry to its history and economy. The town is filled with the historic *bodegas* (cellars) of the great sherry houses, reflecting the importance of maturing in its creation. In the rolling countryside outside the town are the vineyards, with their chalky *albariza* soil – blindingly white in the strong, white sunlight – where Palomino, the principal sherry grape, grows (Pedro Ximénez and Moscatel being the other two grape varieties, from which sweet sherries are made). The wine made from these grapes is then transformed into sherry. In England, sherry has something of a genteel image. In its hometown it is the drink of the people, served with matter-of-factness, but also pride, in bars and restaurants. Drinking small, refreshing glasses of chilled Fino while sitting in a cramped bar – packed with Spaniards listening intently to a commanding flamenco singer – a burly figure of a man – sitting just a few feet away – brought home to me how rooted in day-to-day life sherry is in Jerez.

A trip to the Gonzalez Byass winery, founded in 1835, was especially atmospheric. The old, high-ceilinged, stone-built warehouses, filled with huge barrels, are gracious places, their calm belying the complexity of managing the formidable logistics taken to produce fine sherry. Signatures on several of the barrels from the great and the good who have visited the winery over the decades create an intriguing guest book: Orson Welles, Mario Vargas Llosa, Ayrton Senna, Steven Spielberg among the names I notice. It is here that I meet wine-maker Antonio Flores, a courtly, courteous man who talks with clarity and eloquence about the intricate process of creating sherry. His personal connection to Gonzalez Byass runs deep, as he was born above the oldest cellar in the *bodega*, the one chosen by the real-life 'Tio Pepe' as his personal store. Standing in this small cellar, I watch fascinated as Flores deftly pours sherry straight from the cask into a wine glass using a *venencia*, a small cup on a long stem, used to extract sherry without disturbing the *flor*. I am tasting Fino sherry *en rama* – unfiltered, direct 'from the cask'. It is salty, yeasty, fresh and vibrant; 'a wild Tio Pepe', Flores observes. Traditionally, *en rama* could be experienced only by visiting the winery. In recent years, though, Gonzalez Byass has begun bottling unfiltered Fino each year, releasing a limited number of bottles. Flores explains that the process of selecting which of the 22,000 casks to create the *en rama* from begins after the summer, then is narrowed down in the spring. Gonzalez Byass bottles it and releases it in this season, as this is when the *flor* is particularly active. It is a wine at its best soon after bottling, while it retains its lively character.

A leisurely tutored tasting with Flores gives further insight. 'What this unique ageing system gives us is the possibility to produce the same wine year in, year out,' explains Flores. Within this overall consistency, however, what is noticeable is the sheer range of sherry types. We begin with a pale-coloured Tio Pepe, a filtered Fino aged under *flor* for a minimum of 4 years – crisp and elegant and best drunk within a year of bottling. Next is a 12-year-old Amontillado, made by ageing a 4-year-old Fino for a further 8 years, during which time the *flor* dies because of a lack of nutrients, allowing oxidization to take place and creating a darker colour. The side-by-side tasting offers a chance to appreciate the impact of age on flavour. This is particularly striking when we try first a Nectar, made from Pedro Ximénez grapes, aged for an average of 8 years in oak casks, then Noe, the same wine but aged for an average of 30 years. Practically black in colour, it has a rich, complex flavour, with spicy, chocolate-y notes, dessert in liquid form. 'I would like to age as elegantly as sherry,' says Flores, with a charming smile.

JAMÓN, JAMÓN

In the world of charcuterie, dry-cured meats are regarded with special respect. It takes much time and skill to produce products such as Parma ham and Serrano ham, the results prized for their texture and flavour. One of the world's leading producers of dry-cured hams is Spain, where regulations carefully define its production, protecting traditional processes. A noticeable delicacy is *jamón ibérico*, a dry-cured ham produced only in certain regions of Spain from the native Iberian pig. These long-snouted, slender-legged black pigs are an ancient breed whose origins are thought to go back to Neolithic times. In comparison to the modern white pig, used to make Serrano ham, Iberian pigs have smaller litters, yield less meat and mature more slowly.

Ever since the company was founded in 1988, Spanish food importers Brindisa have done much pioneering work to promote Spain's high-quality hams in the UK. A visit to their Borough Market shop to meet their master carver Santiago Martínez gave me an insight into the intricate world of *jamón ibérico*. 'There are four important factors: breed, diet, the environment and ageing,' he tells me. Ibérico hams are produced in the west and south of Spain, where there is an extensive terrain called the *dehesa*, consisting of grasslands lightly forested with *encina* (holm oak) trees. This natural environment plays a major part in creating the very best of the Ibérico hams, as acorns from these trees are an important factor in the pigs' diet. Of the various Ibérico hams, the most highly prized is *jamón ibérico de bellota*, described by Claudia Roden as 'one of the great foods of the world'. The pigs used for these hams are put on the *dehesa* between October and the end of February to fatten them up, a period of time known as the *montanera*. During this period they feed extensively on acorns (*bellota*), doubling their body weight. Acorns are high in oleic acid, a naturally occurring monounsaturated fatty acid, which in turn manifests itself in the fat of *jamón ibérico de bellota*, giving it a particular melting quality and a complex, nutty, sweet flavour. It is both the quantity and quality of this acorn-fed fat that allows these hams to be cured for as long as they are.

The pigs for *jamón ibérico de bellota* are slaughtered when at least 14 months of age, often around 2 years old. At the start of the curing process the legs, which will form the hams, are salted and set aside for a period. During this time osmosis draws out the water from the meat, replacing it with salt, a process that helps prevent the development of harmful bacteria. The salt is then rinsed from the hams and they are rested for 1 or 2 months in cool conditions with high humidity, allowing the salt to penetrate the meat thoroughly. The hams are then hung up in *secaderos*, drying areas, where they undergo natural drying in ambient temperatures for several months. One of the factors that gives these air-dried hams their aroma and flavour is the range of temperatures the hams are subject to, from the heat of summer to the cold of winter. As the weather heats up, the hams sweat, losing a huge amount of their fat, the meat drying in the process. 'The salting is in the winter, the resting is in the spring, the sweating is in the summer,' Martínez explains. It is, he makes clear, a seasonal product, the pattern of production shaped by the natural acorn harvest. In the years when the acorns are particularly high in oleic acid, the hams have an extra richness. From the *secaderos*, the hams are moved to *bodegas* (cellars), offering a controlled environment to mature further, and are aged in total for up to 4 years. Although the process is highly regulated, regional variations during curing, such as the use of more salt at the beginning of the process in the hotter south, result in a range of ham styles, with the hams from the producers farther north being noticeably sweeter.

Standing behind the Brindisa ham counter, the scent of ham is so powerful that it is almost tangible. The hams are mounted on special stands and carefully, skilfully hand-carved. 'When you've taken all that time – 6 years of waiting – you don't want to spoil the ham with a machine that would damage the texture or slice it too thick,' says Santiago with conviction. 'We carve it carefully, evenly, very finely, making sure that you get the fat and the meat. A slice of ham must be a mouthful. It melts in your mouth and you get all the flavour.'

The variety in flavours and textures that Santiago has been describing becomes clear when I taste the different hams. The Serrano ham, while very good, is straightforward, with a chewy sweetness. The Ibérico hams are recognizable by their long, slender, elegant shape. The flesh is distinctively marbled in appearance, reflecting the way that the Iberian pig stores fat intramuscularly. The Ibérico hams, aged from between 26 months and 4 years, are noticeably more aromatic, the flavour lingering noticeably. I end the tasting with Martínez's favourite – a *jamón ibérico de bellota*, made from pure-bred Ibérico pigs. As I hold the piece of *jamón*, the fat begins to melt at the touch of my fingers. The depth of flavour contained in one small sliver of ham is remarkable: sweet and salty at the same time, with a complex savouriness that lingers on. 'It lasts, it lasts, it lasts,' says Martínez with pride.

THE PLEASURES OF PORT

The story of port wine is rooted in a historic relationship between two countries: England and Portugal, allies since the Middle Ages. During the late seventeenth and early eighteenth centuries, when war with France meant that French wines could not be imported, English merchants turned to Portugal, their old trading partner. So strong were these historic trading links that many English merchants had already taken up residence in Portugal, an important factor in the creation of port. Wines from the Douro, a part of Portugal with a long history of viticulture, began making their way to England. Port, named after Oporto, from where it was shipped, was initially not a fortified wine, but the English fondness for strong, sweet wines led to the rise of fortification – the adding of distilled spirit – a practice that became widely adopted by port producers over the course of the nineteenth century.

Nowadays at the Quinta do Noval vineyard in Portugal, port is still produced traditionally, as it has been for generations. With each harvest, the newly picked grapes are placed in large granite *lagares* to be trodden by foot into grape must (the term used for freshly pressed grape juice). The must is then fermented – a process triggered naturally by the wild yeasts on the grape skins – for several hours. Fermentation, however, is only permitted to reach a certain stage, explains Quinta do Noval's managing director, Christian Seely. The process is tracked using an instrument called a refractometer to measure the sugar content (measured in units called Brix) in the must. The point at which around half the natural sugars have fermented is when the must is drawn off and mixed with 77% *eau de vie de raisin* (grape spirit). This stops the fermentation process, as the yeasts no longer function with this ratio of alcohol. 'That's why port is strong, because you've added the *eau de vie.* Port is also sweet because half of the natural sugars never ferment. It's one of the things that make it unique as a wine. It preserves some of the perfumes and flavours of the fruit.'

Ruby, Tawny, Late Bottled Vintage (LBV), Crusted . . . these evocative names are for different types of port wines. The term 'Crusted' refers to the sediment that is created naturally in the wine as it ages. It was to sift out this debris that the practical custom of decanting certain types of port before drinking arose. The rich array of assorted port wines is created by going down one of two different roads in production: ageing in barrel or ageing in bottle.

Good-quality port has a special capacity to age remarkably well. The most highly esteemed port wines are vintage ports, which represent only a fraction of total port wine production. A vintage year is only declared by port-makers when the wine that year is deemed to be particularly fine. The judgement is one requiring discretion, as the producers' reputation is linked to the quality of their vintage ports. For Seely, a sense of whether or not the wine will be vintage comes early in the process of production: 'You normally have a pretty good idea when the grapes come in and you see the must moving in the *lagares.*' Among the indicators that Seely's experienced eye is noting are the level of bubbling, which signals a good ferment, the rate of colour extraction and the ratio of juice to solids. The actual decision to declare a vintage year, however, is made after the wine has been aged in the barrel, within 2 years.

Seely's interest in port is tangible and infectious. He speaks eloquently of his fascination at being able to compare two ports made at Quinta do Noval in the same year but in different ways. The Colheita is aged in wooden barrels. Seven years is the legal minimum, though Quinta do Noval age theirs for 12. Vintage port, in contrast, is aged in bottles. Vintage port, for Seely, always retains something of the red fruit. The Colheita, however, gains not only its characteristic tawny colour from ageing in wood, but takes on the flavour of dried fruits, nuts and liquorice. 'That's part of the fun of port; it's an extraordinarily diverse wine.' Both Colheita and vintage ports have the capacity to last for around 100 years. This longevity is very much part of port's appeal and mystique.

Among port connoisseurs, Quinta do Noval is famous as the producer of a unique and historic vintage port. At the heart of the Quinta do Noval vineyard are two hectares of very special vines: ungrafted Nacional vines, untouched by phylloxera, the disease that ravaged Europe's vineyards in the nineteenth century and led to much grafting of New World wines on to existing Old World stock. Due to the vines' rarity, the only ports ever made from their grapes are vintage ones, highly regarded and much sought after. Seely tells me that wines made from the Nacional vines follow their own pattern, 'marching to their own drum'. What is a vintage year for the Quinta do Noval vineyard will not necessarily be a vintage year for the Nacional, and vice versa. Sometimes, both the Nacional and the rest of the vineyard will have a vintage year. Although both the Nacional and Quinta do Noval vintage ports are made in exactly the same way, the Nacional will always have its own personality. One of the distinguishing characteristics of a Nacional vintage port, Seely declares, is its 'extraordinary, eternal youth compared to other vintage ports from the same year'. Drinking as he does both Quinta do Noval and Nacional of the same vintage, he is uniquely placed to compare the two ports. 'Always the Nacional will be darker, younger, more tannic, with something a bit more wild about it. The Douro is a wild place and the Quinta do Noval Nacional, as its greatest wine, expresses that characteristic.'

THE AGEING OF BALSAMIC

Italy's traditional balsamic vinegar of Modena is a precious condiment, so time-consuming to produce that it is sold for considerable sums, bottled like the best perfume, in small, elegant flasks. Its origins are ancient. The first written reference to it dates back to the eleventh century, when a small barrel of vinegar was recorded as a gift to royalty. The vinegar has long enjoyed aristocratic associations, and has been given – to royalty and others – by the Dukes of Modena for centuries. Vinegar was for a long time used for medicinal purposes such as dressing wounds, and balsamic vinegar was valued for its health-giving properties, thought to offer protection from the plague and regarded as a tonic. Its history is closely linked to that of winemaking in the area, as it is created from the cooked grape must (grape juice) called *saba*, a traditional ingredient in Modena.

It is an intricate process to make traditional balsamic vinegar of Modena. Historically, its production took place within family homes, where it was stored in the attic and used for special occasions. Just as people would lay down wine for a baby, so barrels of traditional balsamic vinegar would be started to mark the arrival of a newborn. Nowadays, in order to protect it and maintain the tradition, the making of the vinegar is a strictly regulated affair, granted a Protected Designation of Origin and controlled by the Consorzio Tutela Aceto Balsamico Tradizionale di Modena (Consortium for the Protection of Traditional Balsamic Vinegar of Modena). Stipulations regarding the area of production, the use of musts from specific grape varieties cultivated locally, and the production and ageing process are all laid down by the Consorzio. The resulting vinegars are only allowed to be sold at two ages: at least 12 years (Affinato) and at least 25 years (Extra Vecchio). An integral part of the ageing process for traditional balsamic vinegar of Modena is that it takes place in small wooden barrels of assorted sizes. Like sherry, the process for this balsamic follows the *solera* system. When a small quantity of mature vinegar is extracted from the smallest barrel, it is replaced by vinegar from the next barrel in the scale, with the largest barrel at the end of the scale replenished by cooked must from the most recent year. As the cooked must ages it ferments and acidifies, acquiring a distinctive dark colour and complex flavour. As the barrels breathe it also evaporates very, very slowly, thickening over the years to take on an almost syrupy texture.

For centuries, this vinegar, while locally treasured, remained little known outside Italy. Its 'discovery' by the wider world in the latter part of the twentieth century led to it gaining an international reputation as a gourmet product used by chefs in elegant restaurants around the globe. In response to this new market for balsamic vinegar came the production of what is simply called balsamic vinegar of Modena. Made from cooked grape must and wine vinegar, this is a mass-produced product, also regulated by the Consorzio and granted a Protected Geographic Indicator (PGI). A period of 60 days' storage in wood is stipulated, but this is too short a period to affect the colour and flavour. In order to give the characteristic dark colour and sweetness of the traditional balsamic, the addition of caramel is permitted. Another recent mass-produced balsamic product is balsamic glaze, which seeks to mimic the texture of the traditional balsamic by the simple means of adding cornflour. Within the broad framework for balsamic vinegar of Modena (PGI), there is a noticeable range of quality. Price is an indicator, here, as is the ingredients list; the best products contain simply cooked grape must and wine vinegar. Certain producers make balsamic vinegar of Modena more carefully and for longer than the 60 days required, in order to offer an accessible, affordable substitute for the traditional version.

The first thing I notice as I arrive at the workplace of balsamic vinegar producer Nero Modena in the province of Modena is the stately, orderly quality of the grounds. The calmness in the atmosphere seems very appropriate to the making of this venerable vinegar. On my way in, I pass their vineyard, whose Trebbiano and Lambrusco grapes will be used to make the must that starts off the process; the sense of place – a deeply rooted quality – strikes me. Their expert vinegar-maker shows me round one of the *acetaia* (vinegar houses), filled with stacks of wooden barrels, in a range of sizes, containing their precious load. Storing the vinegar in the right surroundings is important to the maturing process. 'The barrels need to feel directly the climate. It is really very important to get the heat of the summer and the cold of the winter. An attic is the best place.' As it matures, the sweet cooked must ferments, taking on acidity with time.

The contact time with wood is key in creating the special flavour of traditional balsamic vinegar. While the use of three woods is legally required, here at Nero Modena they use five in order to achieve the complexity they want. Each of them has a part to play: oak, with its tannins, is essential – 'the king of woods' she says simply – juniper adds piquancy and sweetness, chestnut a dark brown colour, and mulberry and cherry give sweetness again. Gesturing to the wooden barrels before us, she becomes animated, talking of them with a noticeable affection and intimacy as if they were members of her family. 'Cherry is fantastic, but the barrels always leak!' she says, laughing ruefully. She is appreciative of the alchemy that ageing in wood achieves. For her, it is the Extra Vecchio, rather than the 12-year-old Affinato, that truly reflects the ageing process: 'It is in the 25-year-old balsamic that you can really taste the different kinds of wood, taste all of their fragrances,' she says with conviction. 'The flavour is distinctive and complex, sweet and sour in perfect balance.' Working in this way on something that takes so many years to produce requires commitment, I observe. 'One raw material and a long, long time, so you must be passionate,' she acknowledges with a smile.

A VINTAGE EXPERIENCE

To enter Berry Bros. & Rudd's shop at number 3, St James's Street, in the select heart of London's West End, is to feel the living past. This is Britain's oldest wine merchants, founded in 1698 to sell wine to the royal family, St James's Palace and Buckingham Palace being handily close by. The shop is little changed structurally since then. Its presence today is signalled – as it has been since its founding – by the gilded 'Sign of the Coffee Mill' hanging outside the shop, a historic reminder of the business's original role in supplying coffee, then an exotic novelty, as well. Standing in the austere, wood-panelled front room of the shop, my eyes are drawn to a huge pair of scales, originally used to weigh sacks of coffee. From the late eighteenth century onwards, they have famously been used to weigh the shop's notable customers, among them Lord Byron and Beau Brummell, their weights ceremoniously recorded in a ledger. The large, cool cellars below the shop, still used to store wine, at one time housed France's Napoleon III, who held secret meetings there while in exile. Suave and immaculately turned out, Richard Veal, head of events and education, leads me to an atmospheric room in the cellars. Here on the table is a small, stately selection of prime vintage wines, ranging in age from an 1870 Château Margaux to a 2000 Château Lafite Rothschild – bottles that command prices ranging from hundreds of pounds to thousands.

Not all wines will age well. Certain elements have to be present in order for a wine to mature successfully: alcohol content, acidity, fruit extract and, in the case of red wines, tannins. 'These things will give the ability to age,' states Veal. 'To be clear, it's balance but also abundance. If you have a lot of them, you have a more age-worthy wine – acidity, alcohol and tannins are preservatives, so they allow a wine to go on for a long time.' On the whole it is red wines that age particularly well, though there are exceptions, such as Riesling wines and Madeira wines. Nature, as well as the wine-maker's abilities, it transpires, plays an important part in creating a great vintage. In certain years, the impact of adverse weather, such as frost or hail, in the early part of the growing season reduces the flowering and therefore the number of grapes produced by the vines. On the plus side, the quality of these grapes is high. Two of the select bottles in front of us, he explains, come from years like these, 1945 and 1961. Another example of environmental stress that can benefit the quality of the grapes is those years when the conditions are very, very dry. 'Like life, tough times create the great ones,' observes Veal.

During the ageing process, a red wine loses colour, depositing pigmented tannins which form sediment, whereas a white wine darkens in colour, due to oxidization – the same process that sees a slice of cut apple turn from white to brown. With time, the wine softens and the different elements within it begin to unify. Whereas during the early phases of a wine's life the various aspects are distinct and apparent, with time these mesh together. The image Veal uses is of a billiard ball: 'Over the decades the wine becomes a whole, well formed and harmonized, a wonderful spherical unit. Everything's in the right place.' In order to age wine well, minimum movement and disturbance is vital. Ideally, a bottle should have only one or two homes. 'If we know that a bottle has literally come from the merchants in Bordeaux and has been put in our cellars 100 years ago, it's got a good chance of being okay when we open it.' A wine that has been moved a lot is far less desirable, hence much wine is traded without being physically moved, changing ownership while still stored in one place. Storing wines for this length of time brings with it risks. The possibility of imperfect closure means that even a properly stored bottle can disappoint.

One of the many areas of expertise practiced by the wine buyers at Berry Bros. is in tasting new wines and judging their capacity to age well or not. Every year a team from the wine merchant visits Bordeaux to taste out of the cask and make judgements. 'And we will be able to tell if this is a fantastic vintage or not.' What they look for, he tells me, is the elements to be present in a well-balanced way. Berry Bros. & Rudd, it becomes apparent, take a long view, working with wines over decades, assessing them at 5-year intervals. The great wines take several years to develop and reach full maturity. 'We could look at a wine that is 20 or 30 years and think that it still looks fresh. It has such a concentrated mass of all those things we want that it's going to take much longer to mature. The great grape vintages take a lot longer to come around. In the meantime, you can enjoy younger vintages that have reached their peak.' As the wines age in their bottles, through the process of sampling and assessing, consensus on the great vintage years is reached within the wine community. This gradual concurrence on quality seems a very civilized process.

There is a time-capsule quality to the bottles before us, tangible links to past eras in alcoholic form. Here, for example, is an 1870 Château Margaux created in the year when France declared war on Prussia. It is an example of a wine made from pre-phylloxera vines. The latter half of the nineteenth century saw the blight phylloxera devastating France's vineyards, with the solution being to graft phylloxera-resistant American vine rootstocks on to the vines. Here, too, is a 1913 Château de Rayne Vigneau, made before the start of World War I. The half-bottle of Veuve Clicquot is from 1921 – with ageing, Veal tells me, it will have become something more akin to sherry than champagne, yet with acidity too.

Sitting in these calm, peaceful cellars at Berry Bros. & Rudd, surrounded by bottles of wine resting in their racks, I am struck by the patience involved, the investment of time, the sense of history and also the eye to the future, waiting for the right year to arrive when these wines will be at their best. What, I wonder, is it like to sample one of these prized great vintage wines? Veal becomes distinctly animated as he answers, taking me through the experience. The first taste offers a silky texture and a rich concentration of flavour. As you swallow it, even more flavours begin to become apparent. 'The key thing is complexity, especially after you've just swallowed it and it just blossoms. It's very exciting.' Veal is adamant about the uniqueness of what has been created over the years. 'What we're talking about is something here that you can't replicate any other way than by storing a bottle very, very well over a long time. And then you've delivered something which is one of life's great experiences.'

TIME TRAVEL

Food's ability to evoke memories, to transport us back in time, is part of its power and one reason for our fascination with it. While eating is essential for sustaining life, food is so much more than fuel. It is resonant, laden with recollections, emotions, attachments; able to vividly conjure up other times and places. The novelist Marcel Proust's sensuously written passage describing the effect of a madeleine dipped in a cup of tea is the most famous literary example of this: 'No sooner had the warm liquid mixed with the crumbs touched my palate than a shudder ran through me and I stopped, intent upon the extraordinary thing that was happening to me. An exquisite pleasure had invaded my senses, something isolated, detached, with no suggestions of its origin.'

There are good biological reasons for food's capacity to transport us through time and space. Our ability to taste flavours in food draws on our sense of smell. Significantly, the olfactory nerves, which process aroma messages, connect to the amygdalae, areas of our brain that are involved in our emotional responses, including pleasure, fear and anger. The amygdalae are part of the limbic system of nerves and networks in the brain that control emotions and drives, found close to the hippocampus, which plays an important part in creating memories from experiences.

Of all foods, it is the ones encountered when young that seem to have a special power. With a love of sugar hard-wired into us, the sweets we enjoyed as children have particular resonance, enjoyed into adulthood not simply because we like them but because of the happy associations they bring. This level of affection brings with it commercial benefits for much-loved brands. Adults often loyally buy the same confectionery they ate as youngsters throughout their lifetimes. The outrage generated when a brand has the temerity to change a recipe shows the level of emotion involved. There was a furore created in 2015 over Cadbury's decision to replace the Cadbury's Dairy Milk shell on Cadbury's Creme Eggs with a standard cocoa mix. The power of the nostalgic consumer, however, was best demonstrated by Coca Cola's volte face in 1985, when, having introduced a 'new' recipe Coca Cola, they backed down in the face of a massive backlash, bringing back the old Coke and admitting they'd made a major mistake.

Memory and food fascinate novelist Elanor Dymott, who often writes about both in her books and stories. She herself feels the special capacity of food to evoke the past. The mince pies made by her grandmother and her mother have a special place in her affections. Due to her father's work, Dymott lived abroad as a child; her recollections of Christmas in England are limited but vivid. 'We would drive all the way from London to my grandparents' house in Dorset and they'd always have ready this tray of mince pies from the oven.' Dymott's parents, in turn, reciprocated, baking mince pies when her grandparents visited on Christmas Day. 'Our family had very few rituals, but this was one and it stuck in my mind. It was very much about arriving after this long journey, being enveloped in the warm scent of mince pies. They represented the end of a journey, some kind of homecoming, a family embrace.' Last Christmas Dymott set out to make mince pies for herself for the first time, her mother sending her Delia's recipe, her old pastry mixer and a mince pie tin. Making them reminded her of how she had helped her mother do this as a child. Using the very same hand-held tool to mix the lard and butter into the flour made it 'even more of a journey in memory'. Rolling out the pastry brought back the image of her mother doing the same: 'My mother was so quick. One fluid movement of rolling, turning, lifting, turning. Then that lovely thing where you cut out the pastry and lift it up and all those little circles are left there.' As she made them, 'another layer of memory came back with each batch'. She remembered the careful way in which her mother and grandmother had used every last scrap of pastry: 'They took a great pleasure in making as many mince pies as possible from one ball of pastry.' The most vivid is of her grandfather, who had arrived agitated on Christmas Day having had a small accident in his car. 'He was at the kitchen table being upset, then he was no longer being upset because he was presented with a big, steaming mug of coffee and a plate of warm mince pies. The really evocative thing was the power of this food to give comfort and solace.' Both the making and the eating of the mince pies transported Dymott. 'Food is such a powerful way of passing on parts of the past,' she remarks reflectively. 'It is an amazing thing that you can recreate the past through making food.'

The emotional heft of food is one which is being consciously explored by a number of contemporary chefs, seeking imaginatively to intensify the experience they offer diners. At the world-famous El Celler de Can Roca restaurant in Girona, Spain, chef Jordi Roca delights in creating intriguing, playful desserts, using the 'smell of memories' as a starting point for many of his creations. Food's capacity to evoke powerful memories through aroma, texture, taste is similarly enjoyed and explored by British chef Heston Blumenthal, nostalgia being a theme running through his own innovative, famously multi-sensory cooking. When Blumenthal was asked to create space food for British astronaut Tim Peake for his 6-month mission on the International Space Station, the chef's response was to find out what special food memories Peake had, using these as the inspiration for the meals he went on to make. The practical and logistical issues in creating meals that the UK and European Space Agencies and NASA would approve for eating safely in space were considerable, but Blumenthal persevered, delivering a range of tinned meals. Peake's favourite of them all was the salmon creation, flavoured with smoked oil. Eating it, he reported back from outer space, 'takes me right back to Alaska when I was nineteen years old, kayaking around Prince William Sound, catching salmon and cooking it on the fire'. Blumenthal had set out to create food that would connect Peake to Earth, transport him to another time and place, and, amazingly, succeeded.

Food Heritage:

The work of CENTURIES

Our relationship with cooking is a long and profound one. The use of fire to cook food is one of the special traits marking the development of *Homo sapiens*. The ability to grow crops has been seminal to our social development. With the rise of agriculture around 12,000 years ago, farming began to replace hunter-gathering as a means of existence. During the 'Neolithic revolution', people began to grow crops and domesticate animals. Permanent settlements and the ability to feed people resulted in the rise of civilizations. As nations developed over centuries, so, too, did cuisines, composed of culinary customs, the use of specific ingredients in certain ways, and recipes. Our identities are closely bound up with food. In an age of globalization, the way in which people continue to value the idea of national and regional cuisines is striking. These cuisines have been shaped by environmental, historic and social forces over centuries, creating a richness that we take for granted. As seen in this book, the imperative need to eat and to store food for lean times saw human beings develop numerous ways of preserving food. Techniques such as drying, smoking, salting and fermenting ingredients have been used for thousands of years. Everyday foods, consumed without a second thought – bread, yoghurt, cheese – all have an ancient and fascinating history. 'Culture' is a word laden with meaning for fermentation revivalist Sandor Katz: 'We have incredible cultural legacies. Everything about the way that we eat is part of a cultural legacy that we've inherited.'

At a more intimate, personal level, food also passes down within families in the form of recipes, culinary customs, special memories and associations. On a visit to Australia a few years ago, my beloved Aunty Daphne took me to one side to show me how to make pineapple tarts, a traditional Portuguese Eurasian delicacy made for celebrations. A renowned cook within our family circle, she worked by touch rather than measurement, rubbing lard and butter into flour, then mixing in beaten egg to make a soft dough. So soft, in fact, that to shape the tiny tart cases – just 4cm across – she used a special wooden mould. Each case was filled with home-made pineapple jam, then – the hardest part to learn – the edges deftly, neatly pinched using dainty serrated tongs that created a feathered effect. Finally, a small pastry decoration, representing a palm frond, she told me, was cut out and placed in the centre. They take a long time to make, Aunty Daphne acknowledged. 'People ask me why I go to so much trouble. But I don't mind taking the trouble.' In carefully sharing this recipe with me – even giving me my own little mould and tongs for future use – my aunt was passing on her special knowledge, hoping to keep this custom, lovingly practiced by her, alive. Sadly, she died in 2013 and with her passing went a huge amount of knowledge about how to cook dishes that are now no longer widely made, such as *feng* (a historic Eurasian offal curry). When my son turned eighteen, I put together a collection of his favourite recipes and had it printed as a little book for him as a present. Some of the recipes in it are mine, others from favourite cookbooks, still others from his father, grandmothers and dear friends. The fact that he cooks from it regularly gives me real satisfaction.

It is important to recognize, however, that around the world many traditional foods and culinary customs are under pressure, in some cases in danger of being lost altogether. UNESCO's Intangible Cultural Heritage list is designed to focus attention on aspects of culture in order to protect and promote them: many of these are food-related. Kimchi-making in Korea, traditional Mexican cuisine, the Mediterranean diet, gingerbread-making from north Croatia, Hong Kong egg tarts . . . Globalization, industrialization, environmental degradation, economic and social pressures, are among the considerable forces affecting food cultures around the world. The potential loss of food heritage has seen the rise of movements, charities and organizations working to try to preserve it. Some work internationally, such as the Slow Food movement, with its Terra Madre event bringing together thousands of farmers and food producers from 150 countries every 2 years. Others are national, such as the Garden Organic Heritage Seed Library, the Brogdale Collections (home to thousands of rare fruit tree varieties) and the Rare Breeds Survival Trust. In Europe, legal definitions such as Protected Designation of Origin and Protected Geographical Indication have played an important part in protecting historic foodstuffs and traditional production methods. Much, however, has already been lost. 'In every part of the world where we domesticated animals, we developed a really distinctive way of fermenting the milk. Most of those are gone,' says Katz, with both sadness and anger in his voice.

It is not just charities that are important to conserving food traditions. Neal's Yard Dairy, the British cheesemonger's established in 1979, has played a vital role in sustaining Britain's farmhouse traditions, supporting cheese-makers financially and creating a market that made it economically viable for them to survive. Bronwen Percival, Neal's Yard Dairy's cheese buyer, strikes a cautionary note. 'While in many ways we are in a British cheese renaissance, when you scratch beneath the surface a lot of our most classic British cheeses are truly endangered, down to their last single producer.' She cites Kirkham's Lancashire as an example: 'The Kirkhams represent an amazing link between today and what people would have been making 150 years ago, because they have a tactile sense of the cheese-making process and they're controlling it in the same way, bringing out something that's unique. If they shut down we would be left with a Lancashire that bears no resemblance to actual, traditional farmhouse Lancashire.' Percival's earnest desire is to create, in her striking phrase, 'an ecosystem of cheese-makers'. What is needed, she says, is a community of producers, 'not individuals on their own, like pandas in a zoo'.

The last two decades in Britain have seen the rise of new generations of food producers to join the ranks of traditional ones. Many of them are using traditional techniques, though often creating something personal and new, whether a modern cheese or an innovative preserve. Similarly, many people in their own homes are discovering the satisfaction of making their own food, whether nurturing a sourdough, making yoghurt, or turning blackberries into jam. For Katz, teaching people how to ferment food for themselves is key to maintaining this ancient practice. 'All of these foods require people to do them; they can't make themselves,' he says simply. Buying a piece of carefully made traditional cheese or a piece of native breed meat, planting a heritage variety of a vegetable, having a go at making pickles, cordials and chutneys – these are all actions that can help sustain our food heritage. It is sobering to think that – in an age of microwaves and ready-meals – even the simple act of taking the time and trouble to cook a meal from raw ingredients has become an important way of sustaining a food culture passed down to us over the centuries.

Select Bibliography

DAVIDSON, ALAN, *The Oxford Companion to Food*, Oxford, Oxford University Press, 1999.

DEL CONTE, ANNA, *Gastronomy of Italy*, London, Pavilion Books, 2013.

DONNELLY, CATHERINE (ed.), *The Oxford Companion to Cheese*, Oxford, Oxford University Press, 2016.

FEARNLEY-WHITTINGSTALL, HUGH, *The River Cottage Meat Book*, London, Hodder & Stoughton, 2004.

GRIGSON, JANE, *Charcuterie and French Pork Cookery*, London, Grub Street, 2001.

HENRY, DIANA, *Salt Sugar Smoke*, London, Mitchell Beazley, 2012.

HOFFMANN, JAMES, *The World Atlas of Coffee*, London, Mitchell Beazley, 2014.

KATZ, SANDOR, *The Art of Fermentation*, Vermont, Chelsea Green Publishing, 2012.

LIDDELL, CAROLINE, and WEIR, ROBIN, *Ices*, London, Grub Street, 1995.

LOCATELLI, GIORGIO, *Made in Italy*, London, Fourth Estate, 2006.

LÓPEZ-ALT, J. KENJI, *The Food Lab*, New York, W. W. Norton, 2015.

MCGEE, HAROLD, *McGee on Food & Cooking*, London, Hodder & Stoughton, 2004.

MAIR, VICTOR H., AND HOH, ERLING, *The True History of Tea*, London, Thames & Hudson, 2009.

MIODOWNIK, MARK, *Stuff Matters*, London, Penguin Books, 2014.

RANCE, PATRICK, *The Great British Cheese Book*, London, Macmillan, 1983.

ROVELLI, CARLO, *Seven Brief Lessons on Physics*, London, Penguin Books, 2014.

SMITH, ANDREW F. (Editor), *The Oxford Companion to Food and Drink*, Oxford University Press, 2007.

UMAMI INFORMATION CENTRE, UMAMI, Tokyo, 2014.

WEINZBERG, ARI, *Zingerman's Guide to Good Eating*, Boston, Houghton Mifflin, 2003.

WHITLEY, ANDREW, *Bread Matters*, London, Fourth Estate, 2006.

WHITLEY, ANDREW, *Do Sourdough*, London, The Do Book Company, 2014.

WILSON, BEE, *First Bite*, London, Fourth Estate, 2015.

WRANGHAM, RICHARD, *Catching Fire*, London, Profile Books, 2009.

Food and Drink Directory

POTTED SHRIMPS
Baxter's
www.baxterspottedshrimps.co.uk

WINE
Berry Bros. & Rudd
www.bbr.com

FISH COOKERY
Billingsgate Seafood Training School
www.seafoodtraining.org

SPANISH HAMS
Brindisa
www.brindisa.com

SINGLE MALT SCOTCH WHISKY
Bruichladdich
www.bruichladdich.com

MEAT
The Butchery Ltd
www.thebutcheryltd.com

BREAD
The Cake Shop Bakery
www.cakeshopbakery.co.uk

CEVICHE RESTAURANT
Ceviche
www.cevicheuk.com

RESTAURANT, SERVING FORAGED INGREDIENTS
Le Champignon Sauvage
www.lechampignonsauvage.co.uk

CHARCUTERIE
Charcutier Ltd
www.charcutier.co.uk

FISH
The Chelsea Fishmonger
www.thechelseafishmonger.co.uk

SOCCA
Chez Pipo
www.chezpipo.fr

CHOCOLATE
Cocoa Runners
www.cocoarunners.com

KITCHENWARE
David Mellor
www.davidmellordesign.com

JAM
England Preserves
www.englandpreserves.co.uk

CHOCOLATE
Friis-Holm
www.friis-holm.dk

MANGOES
Fruity Fresh
www.fruityfresh.com

SHERRY
Gonzalez Byass
www.gonzalezbyass.com

SEA SALT
Halen Môn
www.halenmon.com

PARMIGIANO-REGGIANO
The Ham and Cheese Company
www.thehamandcheeseco.com

STEAK RESTAURANT
Hawksmoor
www.thehawksmoor.com

HERBAL INFUSIONS AND BITTERS
The Herball
www.theherball.com

PULSES
Hodmedod
www.hodmedods.co.uk

CHOCOLATE
International Institute of Chocolate and Cacao Tasting
www.chocolatetastinginstitute.org

OLIVE OIL CLASSES
Judy Ridgway
www.oliveoil.org.uk

HEREFORD BEEF
Keith Siddorn
Meadow Bank Farm
Whitchurch Road
Broxton
Cheshire CH3 9JS

TURKEYS
Kelly Turkeys
www.kellyturkeys.co.uk

ICES
La Grotta Ices
www.lagrottatumblr.com

MEAT
C. Lidgate Butcher
www.lidgates.com

BARBECUE
Louie Mueller Barbecue
www.louiemuellerbarbecue.com

MAPLE SYRUP
Morse Farm Maple Sugarworks
www.morsefarm.com

FRESH PRODUCE
Natoora
www.natoora.co.uk

CHEESE
Neal's Yard Dairy
www.nealsyarddairy.co.uk

BALSAMIC VINEGAR
Nero Modena
www.neromodena.it

BROWNIES
Outsider Tart
www.outsidertart.com

CHICKENS
Pipers Farm
www.pipersfarm.com

TEA
Postcard Teas
www.postcardteas.com

CLAIRE CLARK – PASTRIES
Pretty Sweet
www.prettysweet.london

PORT
Quinta do Noval
www.quintadonoval.com

HONEY
Regents Park Honey
www.purefood.co.uk

PRESERVES
Rosebud Preserves
www.rosebudpreserves.co.uk

PASTA
Rustichella d'Abruzzo
www.rustichella.it

SMOKED SALMON
Severn & Wye
www.severnandwye.co.uk

COFFEE
Square Mile Coffee
www.squaremilecoffee.com

BARBECUE RESTAURANT
Temper
www.temperrestaurant.com

FRESH WASABI
The Wasabi Company
www.thewasabicompany.co.uk

BEER
Welbeck Abbey Brewery
www.welbeckabbeybrewery.co.uk

RESTAURANT, SERVING OWN-MADE SOURDOUGH
The West House
www.thewesthouserestaurant.co.uk

WHISKY
The Whisky Exchange
www.thewhiskyexchange.com

ASPARAGUS
Wykham Park Farm
www.wykhampark.co.uk

Acknowledgements

This book is a long-cherished project of mine, which has – appropriately – taken years to come to fruition. All books are collaborations and there are several people to whom I am very grateful.

Thank you to my agent, Suresh Ariaratnam, for believing in the idea for this book and for the constant, discerning support as I wrote it. Huge thanks to all the team at Particular Books, including Claire Mason and Shoaib Rokadiya. Special thanks to my editors Cecilia Stein and Helen Conford for their careful, thoughtful work on *The Missing Ingredient*; it's been a real pleasure to work with such committed editors. Thank you, also, to Annie Lee for her meticulous copyediting and editorial manager Rebecca Lee for all her work on the book and her kind, helpful support. I'm very grateful to the design team for making my book look so elegant. Many thanks for their work in spreading the word to Pen Vogler, deputy publicity director, and Olivia Anderson in marketing.

Researching this book was a fascinating and enriching process. I'm very grateful to everyone whom I met, talked to and corresponded with regarding *The Missing Ingredient* for so kindly sharing their time and their considerable knowledge with me.

Many people offered a range of practical help, advice and guidance. My thanks to Angela and Gerry Aroozoo, Russ Carrington, Callum Edge, Helene Cuff, Jessica Goodman, Toby Hampton, Charlie Hicks, Jason Hinds, Mark Lewis, Pam Lloyd, Liz Lock, Erica Marcus, Polly Robinson, Luisa Welch, Patti Wheelan and my son Ben Windsor. Thank you,

Gonzalez Byass and Rustichella d'Abruzzo for your wonderful hospitality.

At a personal level, the support and interest of good friends throughout has been very sustaining. A special thank-you to Cat Black, Tim d'Offay, Geoff Duffield, Joanna Everard, Zoe Hewetson, Nicola Lando, Polly Russell, Helen Smith, Helen Wallace and Lola Wiehahn. A big thank-you to my mother, Lydia Linford, for her meticulous eye and kind encouragement. Last, but not least, heartfelt thanks to my beloved husband, Chris Windsor, for his long, loving support of my work as a food writer.

Index

A

ACTON, ELIZA, 128
ADRIÁ, FERRAN, 105, 173
ALLOTMENT, 53–54
 Noon, Steve, 53–54
AMERICAN CUISINE, 58, 79, 83, 125–26, 174, 182, 226, 304
ARTUSI, PELLIGRINO, 101
ASPARAGUS, 298–300
AUBERGINE, 194, 195–96

B

BAIN-MARIE, 152–54
BALSAMIC VINEGAR (ACETO BALSAMICO TRADIZIONALE DI MODENA), 4, 339–42
 acetaia, 341
 balsamic vinegar of Modena, 339–41
 Nero Modena, 341–42
 solera system, 340, 341–42
 use of wood in creating, 341–42
BARBECUE, 178–88
 brisket, 184–86
 Central Texas, 184–85
 Louie Mueller Barbecue, 183–88
 Meatopia 180–81
BEEF, 83–87, 151, 175, 255, 258–60, 311, 328
 cattle farming, 311–15
 Gloucester cattle, 327
 roast beef, 150–52, 311
 slow cooking, 176–78
 steak, 83–87, 256–57
 traditional Herefords, 312–13
BEER, 250–54
 camra, 251
 Monk, Claire, 252–54
 Protz, Roger, 251–52, 254
 Real ale, 251, 252–254
 Welbeck Abbey Brewery, 252–54
BENNETT, JOE, 267–68
BERRY BROS. & RUDD, 342–46
 Veal, Richard, 342–46
BILTOFT, ELSPETH, 197–98
BLANCHING, 31–32, 195
BLUMENTHAL, HESTON, 25, 93, 123–24, 173, 229, 349

BLYTHMAN, JOANNA, 209
BOCUSE, PAUL, 60
BOTTURA, MASSIMO, 322
BRAISING, 141
BREAD, 143–50
Cake Shop Bakery, 145–50
Chorleywood process, 7–8, 144–45, 147
Real Bread Campaign, 170
sourdough, 143, 169–73
used to test temperature, 92
Wright, David, 145–50
yeast, 143, 146–47, 169, 220
BRILLAT-SAVARIN, JEAN-ANTHELME, 54, 88, 91
BRINDISA, 333–36
BRITISH CUISINE, 63, 69, 78, 104, 124, 136–38, 150–52, 157–60
BROCK, SEAN, 234
BROGDALE COLLECTION, 4, 355
BROWNIES, 125–26
BURNT FOOD, 105–7
BUTCHERS, 84, 87, 257–58, 263, 327–29
BUTCHERY, THE, 258–60

C

CARAMEL, 26–30
CARAMELIZATION 27–30
CEVICHE, 162–63
CHANG, DAVID, 62
CHARCUTERIE, 216–20
bacon, 219–20
British, 217–20
Charcutier Ltd, 217–20
Dunsford, Illtud Lyr, 217–20
ham, 220
jamón ibérico, 319, 333–36
lardo, 286
salamis, 217, 218, 219, 220
CHEESE
affinage, 5, 264–69
Bacterium Linens, 268
Bennett, Joe, 267
blue cheeses, 244–45
cheesemongers, 243, 247
cheddar, 245–46, 265–66
emmentaler, 245
geotrichum, 267–68
Gloucester cheeses, 327
Highfields Farm Dairy, 267
Hodgson, Randolph, 248, 266
Innes Logs, 268–69
Kirkham, Graham, 248–50, 269, 335, 356
Lancashire, 248–50, 269, 355

long finish, 87
making, 9, 241, 243–250
Martell, Charles, 327
maturing, 5, 264–69
milk, 243–50
Montgomery, Jamie, 245–46, 265–66
Neal's Yard Dairy, 243, 248, 265–69, 355
Parmesan (Parmigiano-Reggiano), 24, 25, 321, 322–24
Percival, Bronwen, 265–66, 355–56
Phage, 246
Rance, Patrick, 247–48
Riseley, 268
Stewart, Sarah, 267–69
stored in fridges, 207
St Cera, 269
St Jude, 269
CHEFS
Adrià, Ferran, 105, 173
Arguinzoniz, Victor, 181
Blumenthal, Heston, 25, 93, 123–4, 173, 229, 349
Bottura, Massimo, 322
Chang, David, 62
Ekstedt, Niklaus, 181
Everitt-Matthias, David, 293–94
Garrett, Graham, 171–73
Goin, Suzanne, 116
Henderson, Fergus, 109
Hom, Ken, 80–83
Keller, Thomas, 173
Kinch, David, 25, 61, 236, 237
Koffmann, Pierre, 63, 167–70, 175, 291
Kwong, Kylie, 165–66
Lee, Corey, 62
Nilsson, Magnus, 181,
Oliver, Jamie, 178
Ottolenghi, Yotam, 106, 162
Passard, Alain, 61
Rankin, Neil, 176–77, 178
Redzepi, Rene, 53, 105–6, 195, 293
Roca, Jordi, 349
Sykes, Rosie, 108–10
Turner, Richard, 85–87, 180–81
CHICKEN, 260–62
Greig, Peter, 260–62
Pipers Farm, 260–62
CHILD, JULIA, 60, 107, 108
CHINESE CUISINE, 25, 60, 62, 81–83, 94–95, 103–4, 111, 151, 153, 165–66
CHOCOLATE, 16–23
Christy, Martin, 55–57
conching, 17
Cocoa Runners, 18
Friis-Holm, Mikkel, 19–23
Hyman, Spencer, 18–19
tasting course, 55–57

Young, Paul A., 27
CHORLEYWOOD PROCESS (CBP), 7–8, 144–45, 147
CHOWDRY, NEIL, 297–298
CHRISTMAS, 105, 152, 199, 245, 262, 292, 348
CHRISTY, MARTIN, 55–57
CLARK, CLAIRE, 27–30
COFFEE, 33–43
brewing, 41–42
caffeine, 34
espresso, 35, 38, 40, 41, 42
filter coffee, 42, 43
Hoffmann, James, 36–43
roasting, 37–40
Square Mile Coffee Roasters, 36–43
COLEGRAVE, LIZZIE, 298–300
COOK, RICHARD, 200–04
COOKBOOKS, 127–30
Acton, Eliza, *Modern Cookery for Private Families*, 128
Beck, Bertholle and Child, *Mastering the Art of French Cooking*, 60, 107
David, Elizabeth, *French Provincial Cooking*, 61, 108
Glasse, Hannah, *The Art of Cookery Made Plain and Easy*, 128
Grigson, Jane, *Charcuterie and French Pork Cookery*, 153, 216–17
Liddell, Caroline and Weir, Robin, *Ices*, 307
McGee, Harold, *McGee on Food & Cooking*, 233
Morash, Marian, *Victory Garden Cookbook*, 236
Moss, Robert F., *Barbecue: the History of an American Institution*, 182
Presilla, Marciel, *Gran Cocina Latina,* 230
Pomiane, Edouard de, *Cooking in Ten Minutes*, 130
Shaida, Margaret, *The Legendary Cooking of Persia*, 120
Slater, Nigel, *Real Fast Food*, 130
Smith, Delia, *How to Cook Book One*, 58
Wrangham, Richard, *Catching Fire*, 178
Visser, Margaret, *Much Depends on Dinner*, 235
timings within recipes, 127–30
CORBIN, PAM, 132–33, 138
CORN, 229–237
Brock, Sean, 234
elotado, 232
genetic modification of, 236
grits, 234

Kinch, David, 236, 237
Mennonite farming, 237
nixtamalization, 106, 231
polenta, 233
Presilla, Marciel, 230, 231, 232–33, 235
types of, 229–230
Zea mays, 229
CRACKNELL, SKY, 133–38
CURRY, 112–13, 117–18, 354

D

DALLY, RANU, 193
DASHI, 164, 166
DAVID, ELIZABETH, 60, 61, 108, 157
DAVID MELLOR, 156–57
Mellor, Corin, 156–57
DEEP FRYING, 92–93
DEL CONTE, ANNA, 122–3, 321
DIM SUM, 104
DINING OUT, 7
D'OFFAY, TIMOTHY, 73–77
DRY-AGEING (SEE HANGING)
DUMPLINGS, 104
Dunsford, Illtud Lyr, 217–20
DYMOTT, ELEANOR, 347–48

E

EDGE, CALLUM, 315
EGGS, 58–62
century eggs (*pidan*), 61, 62
custard, 103, 152, 309–10
freshness, 59
omelette, 60
overcooking, 59
meringues, 60
poaching, 60
scrambled, 68
quail eggs, 62
EID, 212
EKSTEDT, NIKLAUS, 181
ELDERFLOWER CORDIAL, 288
ELLIOT, JOHN, 322–24
ENGLAND PRESERVES, 133–38
ESCOFFIER, AUGUSTE, 60, 166–67
EVERITT-MATTHIAS, DAVID, 293–94

F

FARMING AND FARMS, 217–18, 245, 247, 248, 250, 312–314, 353, 355
asparagus farming, 298–300
cacao farming, 56, 57
cattle farming, 84, 87, 259, 311–15
chicken farming, 260–62

dairy farming, 243, 247–48, 265–66, 313, 314
farm to table movement, 53
fish farming, 204
fruit farms, 135
Manor Farm, 265
Meadow Bank Farm, 312–14
Pipers Farm, 261, 262
tea farming, 73, 75
turkey farming, 262–64
wasabi farming, 90–91
Wykham Park Farm, 298
FAST FOOD, 77–80, 92
McDonald's, 79, 92, 93
FASTING, 209–12
FAT, 259–60, 285–88, 313
butter, 285, 286, 287
confit, 287
ghee, 285–86
jamón, 333, 334
lard, 286
lardo, 286
McLagan, Jennifer, 288
potting, 286–87
FAVA BEANS, 158, 159
FEARNLEY-WHITTINGSTALL, HUGH, 101
FEASTING, 212–16
FERMENTATION, 10, 19, 220–24
Katz, Sandor, 222, 224, 353, 355, 356
kombucha, 224
sauerkraut, 221, 222
water kefir, 222–24
FISH
Billingsgate Seafood Training School, 64
cooking, 62–66
fishmongers, 51–52
fresh fish, 50, 51, 52
Goldsmith, Rex, 51–52
Jackson, C. J., 64–66
fasting, 225
FLORES, ANTONIO, 331–33
FOOD HERITAGE, 353–56
FORAGING, 293–94
FREEZING, 135, 204, 221, 255, 274, 303–311
Birdseye, Clarence, 305
domestic freezers, 305–06
ice cream, 306–11
La Grotta Ices, 308–11
Travers, Kitty, 308–11
FRENCH CUISINE, 25, 51, 60, 61, 62, 63, 107, 111, 114–16, 130, 166–67, 168, 170–73
FRENCH FRIES, 92–93
FRENCH LAUNDRY, THE, 30, 173
FRESHNESS, 3–4, 49–54, 207–09
FRUIT, 4, 132, 133, 135, 136, 137, 138, 291, 293, 295
candied fruit, 198–99
fruity fresh, 296–98
Alphonso mangoes, 296–98
FRYING (SEE ALSO

STIR-FRYING), 3, 65, 91–93, 102, 111–14, 116, 117, 122, 128, 177, 195, 219
deep-frying, 62, 130
FUBINI, FRANCO, 295–96

G

GARDEN ORGANIC HERITAGE SEED LIBRARY, 355
GARRETT, GRAHAM, 171–73
GOIN, SUZANNE, 116
GOLDSMITH, REX, 51–52
GREIG, PETER, 260–62
GRIGSON, JANE, 153, 217
GULATI, ROOPA, 117, 121
GUNDERS, DANA, 208–9

H

HAM, 218–220
dry-cured hams, 333–36
jamón ibérico, 4, 330, 333–36
long finish, 87
HAM AND CHEESE COMPANY, 322–24
HAMPTON COURT PALACE, 151, 214–16, 292
HANGING, 255–60
HANNETT, ADAM, 326–27
HAZAN, MARCELLA, 97, 112, 175
HENDERSON, FERGUS, 109
HERBS, 3, 49, 53, 112, 153, 161, 162, 163, 167, 171, 194, 219, 250, 286, 289–91
HERDWICK SHEEP, 328–29
HODMEDOD, 158–60
HODGSON, RANDOLPH, 248, 266
HOFFMAN, JAMES, 36–43
HONEY, 30, 56,198,199, 281–85, 288
beekeeping, 281, 282–84
McCallum, Brian, 282–85
Regent's Park Honey, 282–85
HOM, KEN, 80–83
HYMAN, SPENCER, 18–19

I

ICE, 205, 303–05
ICE CREAM, 307–11
INDIAN CUISINE, 62, 78, 117–18, 120–21, 161, 165, 194, 195, 221, 285–86
INTERMITTENT FASTING, 211

J

JACKSON, C. J., 64–66
JAFFREY, MADHUR, 162
JAM, 105, 131–38, 356
Corbin, Pam, 132–33, 138
England Preserves, 133–38
pectin, 130, 132, 133, 136, 138
sugar content, 132, 133, 137, 138
JAMÓN IBÉRICO, 4, 330, 333–336
Brindisa, 333–36
dehesa, 333
Martínez, Santiago, 333–336
JAPANESE CUISINE, 25, 51, 68, 89, 103, 164, 172,
JEWISH CUISINE, 176

K

KATZ, SANDOR, 222, 224, 353, 355, 356
KELLY, PAUL, 263–64
KIRKHAM, GRAHAM, 248–50, 269, 335, 356
KITCHENWARE, 154, 156
KOBE BEEF, 83–84
KOFFMANN, PIERRE, 63, 167–768, 291
KOMBUCHA, 10
KNUTSEN, KAI, 133–38
KWONG, KYLIE, 165

L

LAMB, 328–29
LANDO, NICOLA, 175
LARDERS, 205
LEA-WILSON, ANDREA, 228–29
LEE, COREY, 62
LEPARD, DAN, 103
LIDGATE, DANNY, 84, 87
LITTLE HOUSE IN THE BIG WOODS, 270
LOCATELLI, GIORGIO, 97
LOPEZ-ALT, J., KENJI, 162, 197
LOUIE MUELLER BARBECUE, 183–88
LUARD, ELIZABETH, 161
LUNGHI, PIETRO, 234
'LA POLENTA', 234

M

MAILLARD REACTION, 3, 35, 67, 86, 102, 116–18, 123, 155, 164, 182
MALAYSIAN CUISINE, 3, 112–14, 196
MALLMAN, FRANCIS, 182
MAPLE SYRUP
Morse, Burr, 271–75
categories of, 274
Grade A, 274
sugaring, 271–75
University of Vermont's

Procter Maple Research Center, 274
MAPLE TREES
black maple (*Acer nigrum*), 271
red maple (*Acer rubrum*), 271
sap of, 271–74
sugar maple (*Acer saccharum*), 271
MARCHETTI, DOMENICA, 195–96
MARINATING, 161–63
MARKETS, 10, 49, 50, 64, 79, 114, 115, 256–257
farmers' markets, 10, 49, 134
MARMALADE, 10, 131–32, 155
MARTELL, CHARLES, 327
MARTÍNEZ, SANTIAGO, 333–336
MAYHEW, HENRY, 78
MAYONNAISE, 107–10
MCGEE, HAROLD, 24, 93, 182, 194, 233
MCLAGAN, JENNIFER, 288
MEADOW BANK FARM, 312–314
MEAT (SEE ALSO BEEF, CHARCUTERIE, LAMB, STEAK), 47, 49, 50, 51, 63, 78, 121, 129, 130, 174, 175–77
barbecue, 178–188
browning of, 68, 106, 158, 154–155
de-chilling, 102
hanging of, 255–60
marinating, 161–63
resting of, 102
roasting, 150–52
stir-frying, 81, 82, 155
wet-ageing, 258
MELLOR, CORIN, 156–57
MELLOWING, 193–94
MELTONVILLE, MARC, 151, 214–215, 292
MICROWAVE, 6, 66–68, 305, 356
MIDDLE EASTERN CUISINE, 78, 119, 120, 161
MILLS, NATHAN, 258–60
MONK, CLAIRE, 252–54
MONTGOMERY, JAMIE, 245–46, 265–66
MORALES, MARTIN, 163
MORASH, MARIAN, 236
MORSE, BURR, 271–75
MOSLEY, MICHAEL, 211
MOSS, ROBERT F., 182–83
MSG, 25–26
MUNIZ, DAVID, 125–26
MUELLER, BOBBY, 183, 187
MUELLER, WAYNE, 184–88
MYHRVOLD, NATHAN, 174

N

NATIVE BREEDS, 259–60, 327–29

NATOORA, 294–96
NEAL'S YARD DAIRY, 243, 248, 265–69, 355
NOON, STEVE, 53–54
NORMAN, JILL, 92

O

OLD, JON, 90–91
OLIVE OIL, 195–96, 301–03
  Ridgway, Judy, 302–03
OLIVER, JAMIE, 178
OLNEY, RICHARD, 60, 193
ORWELL, GEORGE, 74
OTTOLENGHI, YOTAM, 106, 162
OUTSIDER TART, 125–26

P

PAPIN, DENIS, 123
PARMESAN (PARMIGIANO-REGGIANO), 24, 25, 95, 97, 319, 321–24
BOTTURA, MASSIMO, 322
PASTA, 93–101
  rustichella, 97–100, 101
PATISSERIE, 26–30
PATISSIER, 27–30
PAUSES, 101–3
PEDUZZI, GIANLUIGI, 97–99
PERCIVAL, BRONWEN, 265–66, 355–56
PEREIRA, JEANNE, 112–14
PERUVIAN CUISINE, 162–63
PETRINI, CARLO, 79
PHIPPS, CATHERINE, 124
PICKLES, 194–98
  Acar, 196
  Biltoft, Elspeth, 197–98
  British, 195–97, 294
  Italian, 195–97
POMIANE, EDOUARD DE, 130
PORT, 336–39
  Colheita, 338
  Nacional vines, 338–39
  Quinta do Noval, 336–39
  Seely, Christopher, 336–39
  vintage port, 337, 338
PRALUS, GEORGE, 173–74
PRESERVING (SEE ALSO CHARCUTERIE, CHEESE, FAT), 157, 191, 279, 301, 319, 353
  chilling, 199–201
  fermentation, 214–18
  freezing, 283–91
  jams and marmalade, 131–38
  pickling 194–96,
  salt, 224–26
  sugar 198–99
  smoking 200–04
PRESSURE COOKER, 123–24

PRESILLA, MARCIEL, 230, 231, 232–33
PRINCE, THANE, 155–56
PROTZ, ROGER, 251–52, 254
PULSES, 157–60
hodmedod, 158–60
Saltmarsh, Nick, 158–60
soaking, 160

Q

QUINTA DO NOVAL, 336–39

R

RAMADAN, 211–12
RANCE, PATRICK, 247–48
RARE BREEDS SURVIVAL TRUST, 313, 327–29, 355
READY MEALS, 6, 356
REDZEPI, RENÉ, 53, 105–6, 194, 293
REFRIGERATION, 205–07, 259, 293
RESTAURANTS, 62, 73, 91, 97, 101, 105, 165, 177, 178, 201, 258, 293, 310, 331, 340
fast food, 77–78
fish, 63, 64
steak, 83, 85–87
table-turning, 7
RICE, 118–23
absorption method, 118
biryani, 121
RISOTTO, 122–23
soaking rice, 120
socarrat, 120
starch content, 119–20
tahdeeg, 120
types of, 119–20, 122
RIDGWAY, JUDY, 302–03
ROASTING, 3, 21, 141, 150–52, 164, 175, 216, 226
coffee, 36–41, 43, 117
ROBINSON, JANCIS, 88–89
RODEN, CLAUDIA, 176, 334
ROGAN, SIMON, 53
ROSEBUD PRESERVES, 197–98
RUSTICHELLA, 97–100, 101

S

SALT (SEE ALSO PICKLING, CHARCUTERIE, SMOKING), 163, 200, 224–29, 314
brining, 195, 226
fleur de sel, 228
Halen Môn, 228–29
Lea-Wilson, Andrea, 228–29
rock salt, 227
salt cod, 210, 225–26
salt mines, 227
salt pans, 227–28

sea salt, 227–28
SALTMARSH, NICK, 158–60
SAVOURING, 54–57
SEASONALITY, 197, 246–47, 279, 291–300
SEELY, CHRISTIAN, 336–39
SETTING POINT, 131–38
SEVERN & WYE SMOKERY, 200–04
SHAIDA, MARGARET, 120
SHARP, ANDREW, 327–29
SHERRY, 319, 329–33
Amontillado, 330, 332
biological ageing, 330
en rama, 332
Fino, 330, 331, 332
flor, 330, 332
Flores, Antonio, 331–33
Gonzalez Byass winery, 331–33
Jerez, 331
oxidative ageing, 330
Pedro Xim.nez, 331, 332–33
solera system, 329, 330
SIDDORN, KEITH, 311–15
SINGAPOREAN CUISINE, 31, 50, 78–79, 97
SLATER, NIGEL, 130, 291
SLOW COOKING, 175–78
SLOW FOOD, 79, 355
SMITH, DELIA, 58, 103, 104, 152, 263
SMOKING FOOD, 200–04
difference between cold smoking and hot smoking, 200
Severn & Wye smoker, 200–04
smoked salmon, 200–04
smoke products, 202
SO, YAN-KIT, 63, 81, 95
SOCCA, 114–16
SOLERA SYSTEM, 329, 330, 340, 341–42
SOTT'OLIO, 195–96, 301
SOURDOUGH, 169–73, 356
starter, 169
SOUS-VIDE, 173–79
SOYER, ALEXIS, 166
SPEED-EATING, 56, 110–11
SPENCE, CHARLES, 88
SQUARE MILE COFFEE ROASTERS, 36–43
STEAK, 83–87, 256–57
STEAM, 103–5
STEEPING, 3, 288–91
Isted, Michael, 289–91
STEWART, SARAH, 267–69
STIR-FRYING, 80–83, 155, 177
Hom, Ken, 80–83
STOCK, 3, 163–69
ichiban dashi, 164
Koffmann, Pierre, on stock, 163–69
stock cube, 166
STREET FOOD, 79, 178
SUGAR, 198–99

in marmalade and jams, 131–32, 133, 137, 138
in ice cream, 307–310
SUPERMARKETS, 10, 49, 50, 96, 147, 148, 170, 177, 201, 209, 256, 257, 263, 281, 298, 323
SYKES, ROSIE, 108–10

T

TAMIMI, SAMI, 162
TASTE, 14–16
TEA, 68–77
afternoon tea, 69
British relationship with, 69
CTC (crush, tear, curl or cut, tear, curl) tea, 71
d'Offay, Timothy, 73–77
tea-bag, 71, 72, 73
THANKSGIVING, 151, 226, 262
TIME TRAVEL THROUGH FOOD, 346–49
Blumenthal, Heston, 349
Dymott, Eleanor, 347–48
Roca, Jordi, 349
TRAVERS, KITTY, 308–11
TURKEY, 262–64
dry-plucking, 263–64
Kelly, Paul, 262–64
TURNER, RICHARD, 85–87, 180–81

U

UMAMI, 23–26, 56, 164, 175, 244, 255, 323
UNIVERSITY OF VERMONT'S PROCTER MAPLE RESEARCH CENTER, 274
USE-BY DATES, 207–9
USMANI, SUMAYYA, 211–13

V

VEAL, RICHARD, 343–46
VINCENT, JOHN, 79–80
VINEGAR (SEE ALSO BALSAMIC VINEGAR), 109, 127, 161, 194, 289
VINTAGE WINE, 342–46
VISSER, MARGARET, 235

W

WASABI, 89–91
Old, Jon, 90–91
Wasabi Company, 90–91
WATERS, ALICE, 53
WHISKY, 324–27
Bruichladdich distillery, 325–27
extra-aged whiskies, 325
Hannett, Adam, 326–27

No Age Statement (NAS), 325
Singh, Sukhinder, 325
single malts, 324
Whisky Exchange, 325
WILDER, LAURA INGALLS, 270
WRANGHAM, RICHARD, 178–79
WRAP, 208
WRIGHT, DAVID, 145–50
WYKHAM PARK FARM, 298–300

Y

YEAST, 143, 169, 220, 221, 252, 253, 254, 330
YOUNG, CHRIS, 170

Z

ZABAGLIONE, 154